RELIABILITY, MAINTAINABILITY AND
SUPPORTABILITY: A PROBABILISTIC APPROACH

RELIABILITY, MAINTAINABILITY AND SUPPORTABILITY: A PROBABILISTIC APPROACH

Jezdimir Knezevic

School of Engineering, University of Exeter

McGRAW-HILL BOOK COMPANY

London · New York · St Louis · San Francisco · Auckland · Bogotá
Caracas · Hamburg · Lisbon · Madrid · Mexico · Milan · Montreal
New Delhi · Panama · Paris · San Juan · São Paulo · Singapore
Sydney · Tokyo · Toronto

Published by
McGRAW-HILL Book Company Europe
Shoppenhangers Road, Maidenhead, Berkshire, SL6 2QL, England
Telephone 0628 23432
Fax 0628 770224

British Library Cataloguing in Publication Data

Knezevic, J.
  Reliability, Maintainability and
  Supportability: Probabilistic Approach
  I. Title
  620.004

  ISBN 0-07-707423-8

Library of Congress Cataloguing-in-Publication Data

Knezevic, J. (Jezdimir),
  Reliability, maintainability and supportability: a probabilistic
  approach / J. Knezevic.
    p. cm.
  Includes index.
  ISBN 0-07-707423-8
  1. Reliability (Engineering)   2. Maintainability (Engineering)
  I. Title.
  TA169.K58   1993
  621'.0045–dc20   93-17578
                                                                   CIP

1234 CUP 9543

Typeset by Cambridge University Press

and printed and bound in Great Britain at the University Press, Cambridge.

To my Professors and Friends:
Joca, Aleksandar Mihailovic, Bob, JOF, Chris, Ben.

CONTENTS

PREFACE

Who starts with certainty ends in doubt.

This book is written by an engineer for engineers. It addresses, through a non-engineering approach, those engineering disciplines which deal with the reliability, maintainability and supportability of man-made products.

Engineering is a scientific discipline in which relationships are governed by rules and laws of a deterministic nature. As a consequence, the measures which describe quantitatively the characteristics of man-made products have specific values. For example, characteristics such as: speed, acceleration, weight, volume, capacity, voltage, length, width, shape, road clearance, size of memory, drag coefficient, conductivity, and many others are single-value numbers for the products under consideration. Thus, engineers are people whose minds are orientated towards a deterministic way of thinking. However, when it is necessary to describe such products from the point of view of their reliability, maintainability and supportability, it becomes clear that the deterministic approach does not work. In order to quantify measures for these particular characteristics it is necessary to apply *probability* as a tool. This causes a problem for engineers because they are not trained to think probabilistically, in that words such as possibly, most likely, probably and similar are not used to describe speed, power, weight, capacity and similar properties.

The intention of this book is not to change engineers' way of thinking, but rather to make them more familiar with real-life situations where a probabilistic way of thinking is essential. Thus, the main aim of this book is to assist engineers when dealing with man-made products whose characteristic measures can only be quantified through the probabilistic approach. The main objectives of this book are therefore to introduce:

(a) The *concepts* of characteristics of a product which define its function-ability profile
(b) The *measures* which describe those characteristics, in qualitative and quantitative form
(c) The methodology for the *prediction* of these characteristics
(d) The methodology for the quantitative *assessment* of the measures of these characteristics, based on the available empirical data.

In order to make the presentation easier, specific meanings are attached to certain words throughout this book. Thus *item* is used as a generic term for a product, module, subsystem, component and similar, when analysed as a single object, and *system* is used as a generic term for products which are analysed as a set, of many consisting objects.

This means that the word item will be used for a car in the same way as for an engine, a carburettor and a distributor cap, when treated as a single object. On the other hand the word system will be used for a car when it is treated as a set of consisting objects such as the engine, transmission, body, brakes, etc.

Functionability, the term which I use to represent the essential character-istic of a product or system, is introduced in Chapter 1. In the same chapter it is pointed out that this characteristic does not define the functionability profile of a product during its operational life. The characteristics which do so are reliability, maintainability and supportability, and these concepts are introduced in Chapters 2, 3 and 4 respectively.

Probability theory is employed as a tool to quantitatively define these characteristics and the concept of the probability system is therefore given in Chapter 5, together with a concise version of the fundamentals of prob-ability theory. It is important to emphasize that this book is not intended as a rigorous treatment of the relevant theorems and proofs. Rather, the intention is to provide an understanding of the main concepts behind prob-ability theory, and to show the practical application of previously derived definitions and expressions to characteristics analysed in everyday engineer-ing practice. A brief description of the theoretical probability distributions which are most frequently used within the engineering disciplines concerned is given in Chapter 6.

The measures which quantitatively define the reliability, maintainability and supportability of an item are defined and derived in Chapters 7, 8 and 9, where a large number of worked examples are used to illustrate their practical application. The concept of the characteristic which establishes the relationship between reliability, maintainability and supportability, known as *availability*, is given in Chapter 10.

In Chapters 11, 12 and 13 methods for the prediction of reliability, maintainability, and supportability measures are derived for a system, com-plex maintenance task and support task, based on corresponding measures

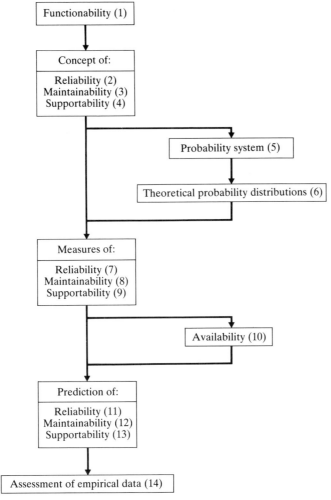

Figure 0.1 Flowchart of the book content.

of the consisting items, tasks and activities. Chapter 14 deals with methodologies for the estimation of reliability, maintainability and supportability measures based on available empirical data.

Intentionally, the structure of chapters related to the same topic (2, 3 and 4, 7, 8 and 9, 11, 12 and 13) is kept as similar as possible, in order to demonstrate the universal application of the probability system to different measures of various characteristics. The structure of the topics covered by this book is also illustrated by the flowchart of the contents in Fig. 0.1.

As a mechanical engineer who has spent over ten years dealing with the application of the probability system to characteristics such as reliability, maintainability and supportability, I would be extremely happy if my experience, as summarized in this book, can be of value to all engineering practitioners and students alike.

J. Knezevic

ACKNOWLEDGEMENTS

I wish to thank all undergraduate and postgraduate students and partici-
pants on short courses from industry who have, through the years, shaped
this text by providing me with the 'feedback' necessary for its continuous
improvement. I would like to extend my sincere thanks to two of my students, Olav Jones and Martin Nethercote, for their contribution to the final
version of the software. I also extend my appreciation to Thelma Filbee for
typing and printing many versions of this book during the last two years,
and to Lynn who patiently coped with all of this and myself.

ONE

THE CONCEPT OF FUNCTIONABILITY

The only commonality between all the man-made products around us (cars, plants, roads, shoes, houses, pencils etc.) is the purpose of their existence which is to *perform a specified function.*

In a sense, each product is the solution to a problem. For example, whenever one wants to make a cup of coffee, there is the problem of heating the water to boiling point. One possible solution to this problem is to use a kettle, a product whose main purpose of existence is to perform that function. It follows that every product which performs a specified function can be said to be functional. Thus, the kettle is a functional man-made product which heats water to boiling point.

A given product is not only expected to perform a specified function but it is also expected to perform in a *satisfactory manner.* Thus a kettle which needs, say, 45 minutes to heat a litre of water to boiling point is not satisfactory. The expression *satisfactory manner* is a common description for the requirements which the product has to satisfy while performing the specified function. Most often these requirements are related to size, weight, volume, shape, capacity, flow rate, speed, performance and many other physical and operational characteristics.

In order to derive a product which performs a required function in a satisfactory manner it is necessary to specify the *operating conditions* under which the product will be used. In the case of a kettle, the operating conditions are primarily related to the mains supply voltage, vibrations, humidity and similar factors. Therefore, the aspects of *functionality, satisfactory manner* and *specific operating conditions* have to be brought together in order to obtain a full picture about the product under consideration.

I shall call the characteristic which embraces all three aspects *functionability*, which is defined as:

The inherent characteristic of a product related to its ability to perform a specified function according to the specified requirements under the specified operating condition.

In the above definition the word 'inherent' is used to stress that all decisions related to the functionability of a product are made at the design phase. For instance, a functionable motor vehicle is one which performs a transportation function, satisfying specific requirements such as speed, fuel and oil consumption, acceleration, load (number of passengers and luggage), height, width and length, riding comfort, and many other features when it is used under specified operating conditions (type of road surface, terrain configuration, outside temperature, fuel grade, etc.). In reality the list will be much longer and more exhaustive. For example, a car which performs in a satisfactory manner on the public roads in England might not be able to do so in the Sahara Desert or on the snow-covered roads of Siberia.

In order to analyse a product or to compare products it is necessary for functionability to be expressed quantitatively. In practice functionability is, in the majority of cases, numerically described through a set of data commonly called *performance data*. On the other hand, there are some characteristics of functionability which unfortunately cannot be numerically expressed and in these cases it is necessary to apply a rating scheme. In both situations it is necessary to specify the operating conditions under which these data are obtainable, so that realistic comparisons can be made.

Example 1.1 In order to illustrate the above statements, consider Table 1.1 which shows the available information on the performance of three different makes of car. Let's call them models X_1, X_2 and X_3. According to their functionability characteristics all three are comparable. It is necessary to say that the data are obtained under the following test conditions:

- Surface: dry tarmacadam
- Temperature: 32°F, 3°C
- Wind: 10 mph
- Barometer: 1019 mbar
- Weight: unladen, no fuel, driver only

Table 1.1 Performance data

		X_1	X_2	X_3
Max speed	mph	101.9	107.8	101.2
Max in 3rd gear	mph	77	88	92
Max in 2nd gear	mph	51	61	61
Max in 1st gear	mph	28	33	35
0–60 mph	sec	10.6	11.4	11.6
30–50 mph in 4th gear	sec	10.8	10.4	10.4
50–70 mph in top gear	sec	10.8	10.4	10.4

Table 1.2 Rating data

	X_1	X_2	X_3
Handling	****	****	***
Ride comfort	****	****	**
Accommodation	****	****	***
At the wheel	****	***	****
Visibility	****	****	****
Instruments	***	****	***
Heating	***	****	****
Ventilation	****	****	****
Noise	****	****	***
Finish	****	****	****

- Fuel grade: 97 octane (4 star rating)

Thus, the performance is obtainable under these conditions, which means that the change in one or more of them will cause the change in performance. Clearly, the degree of change depends on the magnitude of variations of the operating conditions from the nominal values.

As an illustration of the rating approach, Table 1.2 lists non-numerically described characteristics which are defined through the rating system: excellent *****; above average ****; average ***; below average **; poor *. It should be said that the grades obtained from applying the rating method represent the personal view of the assessor. One has to be cautious in dealing with the rating method because it can be very subjective and prejudiced.

1.1 FUNCTIONABILITY PROFILE

The state of a product in which it is functionable will be called the state of functioning, denoted as *SoFu*. Clearly, an item is functionable at the beginning of its operational life but irrespective of how perfect the design of a product may be, or the technology of its production or the materials from which it is made, during its operation certain irreversible changes will occur. These changes are the result of processes such as corrosion, abrasion, accumulation of deformations, distortion, overheating, fatigue, diffusion of one material into another, and similar. Often these processes superimpose on each other; they interact with each other and cause a change in the condition of a product, as a result of which its basic characteristics and performance will change. The deviation of the characteristics of the product from the accepted nominal values or range of values is considered a failure.

Additionally, failures can be caused by factors unconnected with the processes, occurring in the materials of which the product is made. This type of failure is caused by a combination of unfavourable factors. For example, the thermal cracking of a product as a result of a defect in the lubricant or cooling supply, the fracture of a product resulting from incorrect use or overloading, deformation or fracture of a product caused by operating conditions in which each parameter is at an extreme (maximum loads, minimum hardness of material, excessive temperatures, etc.), and the effect of technological defects.

The failure of the product can therefore be defined as an event whose occurrence results in either the loss of ability to perform the required functions or the loss of ability to satisfy the specified requirements.

Regardless of the reasons for its occurrence, a failure will cause the transition of a product from the state of functioning to a new state, known as the state of failure, *SoFa*. For some products transition to the state of failure means retirement, for example rockets, satellites, batteries, light bulbs, resistors, fuses, chips, and bricks. Their functional pattern can be represented as shown in Fig. 1.1. Products of this type are known as non-restorable simply because it is impossible to restore functionability after failure.

Figure 1.1 Functionability profile of a non-restorable product.

Conversely, there are a large number of products whose functionability can be restored; these types of products are known as *restorable*. Thus, when one says that a specific product is restorable, it is understood that after it has failed its functionability can be restored. Consequently, the term *restorability* will be used to describe the ability of the product to be restored after its failure.

In order to restore the functionability of a product it is necessary to perform specified activities known as *maintenance tasks.* The most common restoration activities are cleaning, adjustment, lubrication, painting, calibration, replacement, repair, refurbishment, renewal, and so on; very often it is necessary to perform more than one activity in order to restore the functionability of a product. Apart from the maintenance activities caused by failure during operation, a product may require some activities to be performed, simply to retain it in a state of functioning. Generally speaking these activities are less complex than those needed for the restoration of functionability, and are typified by cleaning, adjustment, checking, and inspection. The set of activities performed in order to restore to, or retain in, *SoFu* is called the *maintenance task.*

It is necessary to stress that maintenance tasks cannot be performed without appropriate resources such as spares, material, trained personnel,

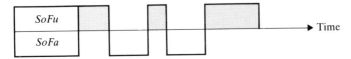

Figure 1.2 Functionability profile of a restorable product.

tools, equipment, manuals, facilities, software, etc. As the main task of these resources is to support the maintenance task we shall call them *maintenance support resources*, MSR. The set of all activities (engineering and administrative) which have to be performed in order to provide all required support resources will be called the *Support Task*.

The view of functionability, during the operational life of a restorable product which fluctuates between *SoFu* and *SoFa* until its retirement is shown in Fig. 1.2. The established pattern is termed the *functionability profile* because it maps states of the product during its operational life cycle. Usually calendar time is used as the unit of operational time against which the profile is plotted.

According to Fig. 1.2, for any given moment of operating time, the product under consideration, from the point of view of functionability, could be in one of the two possible states:

- State of Functioning, *SoFu*
- State of Failure, *SoFa*.

1.2 BEGINNING OF OPERATION

Every engineering product is in *SoFu* at the beginning of its operational life and is fully defined by the information provided by the manufacturer which relates to:

1. *Performance*, which quantitatively describes its functionability;
2. *Technical characteristics*, which quantitatively describe its physical properties;
3. *Price*, which quantitatively describes its acquisition cost.

Some additional information could also be available regarding safety and legal aspects of operation, instructions for use, type and condition of guarantee, and similar features. This information is easily accessible from a variety of catalogues, newspapers, journals and brochures.

Example 1.2 This example represents the continuation of Example 1.1 in that the data shown in Tables 1.3–1.5 are related to the technical characteristics and cost of the same products, motor cars X_1, X_2 and X_3.

Table 1.3 Technical characteristics data

Characteristics	unit	X_1	X_2	X_3
Length	mm	4470	4390	4370
Width	mm	1700	1720	1670
Height	mm	1400	1360	1390
Wheelbase	mm	2570	2610	2570
Front track	mm	1460	1450	1400
Rear track	mm	1470	1460	1400
Weight	Kg	1016	965	955
Drag coefficient		0.37	0.34	0.39
Boot capacity	lit	420	340	400
Engine capacity	cc	1598	1596	1598
Power	KW/rpm	85/5600	90/5400	90/5800
Torque	KN/rpm	/3500	/3500	/3800
Tank capacity	lit	50	60	61
Brakes		disc/drum	disc/drum	disc/drum
Tyres		180/60R365	195/65/14	165SR13
Turning circle	m	10.6	9.9	10.2
Turns lock to lock		4.1	4.2	4.2

Table 1.4 Cost data

Characteristics	X_1	X_2	X_3
Price, inc VAT & tax	7948	7554	7891
Insurance group	4	4	4
Set brake pads	35.65	26.02	23.17
Complete clutch	85.33	90.92	79.81
Complete exhaust	111.69	114.01	126.04
Front wing panel	36.80	52.30	62.68
Oil filter	5.06	2.93	2.96
Starter motor	139.21	35.23	59.80
Windscreen	63.83	35.43	40.07
Battery	43.0	25.70	43.50
Clutch	135.0	94.0	79.55
Alternator	132.0	50.1	61.10

Table 1.5 Maintenance data

		X_1	X_2	X_3
Major service	miles	24000	12000	18000
Intermediate	miles	12000	6000	9000
Major service	h	3.5	2.4	1.2
Change clutch	h	3.5	2.0	1.1
Change water pump	h	1.1	0.9	1.2

Also, there will be a long list of 'eye catching' equipment such as a lockable glovebox, split folding rear seat, map-reading light, boot light, central door-locking, electric mirror adjustment, sun roof, and many more features which define the comfort and ease of use.

It is extremely important for the user to have information about the functionability, cost, safety, and other characteristics of the product under consideration at the beginning of its operational life. It is equally, or even more, important to have information about the characteristics which define the pattern of its functionability profile. This is particularly true with regard to restorable products where there is a need for quantitative information to satisfy questions such as:

- *How long is the product going to be in* SoFu?
- *How long is the maintenance task going to last?*
- *How long is the support task going to last?*

Unfortunately answers to the above and similar questions cannot be found in glossy catalogues, journals or brochures. In the majority of cases the users are left to find the quantitative answers related to the functionability profile, in spite of the fact that they have not been involved with the product during the design stage, when the fundamental decisions have been made, or during the production stage when they have been realized.

Example 1.3 Partial answers to the first question, for the types of cars, X_1, X_2, and X_3 have been generated by the Hertz leasing company and their findings published in an article 'Reliability—the real-life evidence' in the *Sunday Times* on 23 January 1986. Their study was related to the following four items: battery, clutch, alternator and starter motor, and some results are shown in Table 1.6.

Figure 1.3 illustrates the clear division between known and unknown information available at the beginning of the operational life of each product.

The main objectives of this book are to introduce those characteristics of a product which are known as *reliability, maintainability* and *supportability* and which define the functionability profile, together with the measures which quantify them.

Table 1.6 Reliability of several items

Item	mileage	X_1 %	X_2 %	X_3 %
Battery	25 000	14.1	16.6	1.5
	35 000	29.9	22.8	3.9
	45 000	43.2	29.7	5.8
Clutch	25 000	5.3	14.0	0.4
	35 000	15.0	30.0	14.0
	45 000	41.0	62.0	27.0
Alternator	25 000	15.2	2.2	2.3
	35 000	17.9	6.3	3.5
	45 000	25.4	7.7	5.4
Starter Motor	25 000	5.4	1.1	1.1
	35 000	14.1	4.4	2.3
	45 000	26.1	8.5	3.3

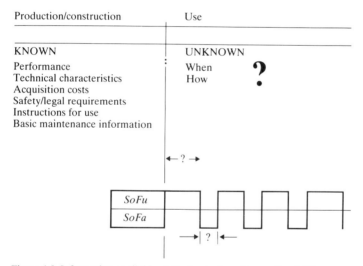

Figure 1.3 Information available at the beginning of operational life.

THE CONCEPT OF RELIABILITY

- How long is the product going to be in *SoFu*?

This question is directly related to the upper part of the functionability profile shown in Fig. 2.1. Instead of providing the information which satisfies this type of question the first page of the manual for the newly acquired product will, in the majority of cases, contain sentences similar to these:

> Congratulations on buying *Xbrand*. We feel certain many years of trouble-free experience will confirm the wisdom of your choice.
> *Xbrand* is intelligently designed for maximum versatility, and superbly constructed to give year after year of trouble-free service.

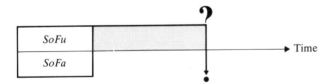

Figure 2.1 Uncertain length of time in *SoFu*.

The best that a producer offers is some sort of guarantee which is valid for a specified interval of use under specified conditions. It is necessary to emphasize that the guarantee does not mean that the product will not fail within a specified period. It only means that if or whenever the product fails they will repair it free of charge. If the user wants to know what to expect from the newly acquired product when the guarantee has expired, the answer could be found in the last paragraph of the user manual which might say:

> After the expiry of the guarantee period, we will willingly service and maintain your product as necessary, at only a modest cost to yourself.

The nature of the information provided by the designer and manufacturer does not help the user to form an idea about the functionability profile of a newly acquired product which is vital for:

1. All decisions which have to be made regarding timing, specifying and planning of maintenance activities as well as all decisions relating to support resources.
2. Analysis of the product to be acquired. If there are two or more products with the same functionability, then naturally the one which will stay in *SoFu* longer will be chosen.
3. The deployment of a new product where the consequences of failure could affect human lives or environment.

Objectively speaking the statement 'years and years of trouble-free service' does not mean a lot to the user, especially today when every producer says the same. The user needs a quantitative description of this and similar phrases. As no scientific discipline was able to help designers and producers in providing satisfactory answers to the above and other related questions, the need arose to form a new discipline and *reliability engineering* was created. This can be defined as:

> A scientific discipline which studies processes, activities and factors related to the ability of a product to maintain functionability during operating time and works out methods for its achievement, prediction, assessment, and improvement.

Reliability engineering is rapidly growing in importance because it is assuming a very significant position in every phase of the life cycle of a product, having impact upon the design, production, operation, maintenance and support tasks. At the same time reliability theory provides a very powerful tool for engineers to quantify the ability of their products to maintain functionability during operational life.

2.1 DEFINITION OF RELIABILITY

In technical literature several definitions for reliability can be found. The list could be increased considerably by including definitions of reliability related to non-technical subjects such as reliable employee, reliable friend, reliable dog and so forth. In this book the following definition is used:

Reliability is the inherent characteristic of an item related to its ability to maintain functionability when used as specified.

To be of practical value reliability, as defined above, has to be expressed numerically. Descriptive statements such as very reliable, quite reliable, year after year of trouble-free service, etc., have to be quantified. That is, the qualitative characteristic has to be translated into a quantitative measure. As a first step it is helpful to try to explain the meaning of reliability. Since there is no item which is able to maintain its functionability for an infinite period of operating time, it is necessary to establish a link between functionability and the length of operating time during which it is sustained. Reliability is graphically presented in Fig. 2.2 where T represents the instant of time when the item lost the ability to function, i.e. transition to *SoFa* occurred.

Functionability is represented by a horizontal line because it is constant while the item is in *SoFu*. It is possible to assign a numerical value to it, but as it would be identical for all items under consideration there is no need here.

Despite the fact that Fig. 2.2 represents only an illustrative attempt at defining the meaning of reliability, it also suggests that the ability of the item to maintain functionability could be numerically expressed by the shaded area. This is quite an acceptable suggestion because reliability is directly proportional to the area, in that the more reliable an item is, the larger the area covered, and vice versa. It is necessary to stress that the size of the area depends mainly on decisions taken during the design phase. Decisions regarding the order of magnitude of the time which the item will be in *SoFu* (5 days, 5 months or 5 years) can only be taken at very early stages of design, through other decisions relating to the selection of the material, shape, technology, surface finish, tolerances, and many other parameters. Thus, reliability can be quantitatively expressed through the length of the operation T during which the item under consideration maintains functionability, when used as specified.

The question which immediately arises here concerns *the nature of the parameter T*. Before trying to provide the answer, it is necessary for the following point to be made. Any given item exists only through its manifestation as an end product. There will generally be many individual copies, each with the same functionability as the item under consideration,

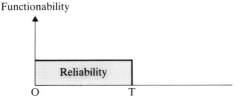

Figure 2.2 The reliability of an engineering item.

Table 2.1 Reliability of gearboxes

14 588	33 000	34 021	50 819	32 381	33 636	28 182
52 880	52 203	28 802	18 527	35 024	44 521	24 702
34 361	22 781	18 118	18 233	23 839	29 133	24 390
28 112	33 427	24 008	30 793	30 449	19 819	47 681
15 642	16 120	14 580	26 999	17 281	40 245	23 960
23 689	22 797	23 805	35 316	15 411	42 490	28 890

but at the same time the reliability of the item will depend on the behaviour of individual copies. Thus, the above question could be rephrased: is T constant for every single copy of the item, or does it vary from copy to copy?

In order to derive some feeling about the behaviour of an item during its operational life, data relating to the occurrence of gearbox failure for a three-door hatchback were collected. The data are shown in Table 2.1 which presents the numerical values in miles which parameter T has taken in the case of 42 copies of the observed product.

In general, if we analyse the operation of several, supposedly identical, copies of an item we expect each of them to perform the specified function under similar operating conditions and regimes. After a certain operating time one of them will lose its ability to maintain functionability, and will reach *SoFa*, which will be denoted by a_1. The other copies will continue operating until the next failure occurs, denoted by a_2, and the process continues until the last copy fails, denoted by a_n, as illustrated by Fig. 2.3.

The example given confirms what everyone familiar with the behaviour

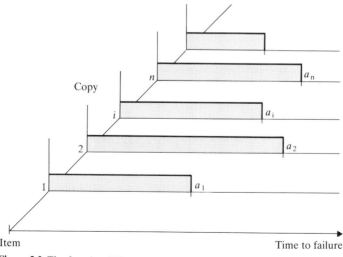

Figure 2.3 The functionability profile of several copies of an engineering item.

of engineering items in practice already knows—that every particular copy of the item under consideration will lose functionability at a different instant of operating time.

The question which naturally arises here is why copies of the item under consideration are failing at different instants of operating time when the same design and the same production process was applied to all of them. In order to find the answer it is necessary to analyse all the contributing factors, of which the following three groups are the most influential:

- *Inherent factors*—those which are related to the strength and condition of the raw materials of which each copy was made, manufacturing and assembling processes, quality control to which each copy was exposed, and initial packaging, transportation and storage of each copy prior to the beginning of operation, and so forth.
- *Environmental factors*—these represent the influence of temperature, humidity, pH concentration, air pollution, dust, terrain, radiation, vibration, etc.
- *Operational factors*—these take into account parameters such as loading, number of cycles per unit, intensity of use, frequency of use, the technical education of users, misuse and storage conditions.

Thus, the variation in instants of time for transition to *SoFa* for each individual copy is the result of the influence of some of the above mentioned factors and the interactions between them. Consequently, the nature of T, for the item under consideration, depends directly on the variability of those parameters, and the relationship between the influential factors and T can be expressed:

$$T = f \ (inherent, \ environmental \ and \ operational \ factors) \qquad (2.1)$$

Analysing the above expression it could be said that as a result of the large number of influential parameters in each group, coupled with their variability, it is practically impossible to find the rule which would deterministically describe the very complex relation denoted by f. Consequently, the parameter T has an unpredictable nature which results from the variability of all the (irreversible) processes occurring in the item materials whilst in use, under the combined influence of the factors mentioned above. Thus, it is impossible to give a deterministic answer regarding the instant of operating time when the transition from *SoFu* to *SoFa* will occur for a given copy of the item.

The only way forward with reliability analysis is to employ *probability theory* which offers a tool for the probabilistic description of the relationship defined by Eq. (2.1). It is then possible to assign a certain probability that the transition will happen at a given instant of operating time, or that a certain percentage of items will or will not fail by a specific instant of time.

CHAPTER
THREE

THE CONCEPT OF MAINTAINABILITY

- How long is the maintenance task going to last?

The above question is directly related to the lower part of the functionability profile, as Fig. 3.1 shows. In the majority of cases the answer cannot be found in glossy catalogues because the main concern of designers is the achievement of functionability. Historically, aspects of restoration are ignored, in spite of the fact that answers to all questions of this nature depend on design.

Figure 3.1 Uncertain length of restoration time.

Today, the situation is gradually changing, thanks to aerospace and military customers who recognize the importance of this type of information and have established it as an important characteristic, equally desirable for performance and reliability. As no established scientific disciplines were able to help designers and producers to provide answers to these problems the need arose to form a new discipline and *maintainability theory* was devised, defined as:

> A scientific discipline which studies the activities, factors and resources related to the restoration of functionability of a product by performing specified maintenance tasks, and works out methods for quantification, achievement, assessment, prediction and improvement of this characteristic.

Maintainability engineering is rapidly growing in importance because of its considerable contribution towards the reduction in maintenance costs of a product during its operation. At the same time, maintainability theory provides a very powerful tool with which to provide engineers with a quantitative description of the ability of their products to be restored to functionable order, by performing appropriate maintenance tasks.

3.1 DEFINITION OF MAINTAINABILITY

In the technical literature several definitions of maintainability can be found. In this book the following definition is used:

> Maintainability is the inherent characteristic of an item related to its ability to be restored when the specified maintenance task is performed as required.

In order to be useful to engineering practice, maintainability, as defined above, has to be expressed numerically. That is, the qualitative characteristic has to be translated into quantitative measures.

In order to explain the physical meaning of maintainability, let us establish the link between the specified maintenance task and the length of time required to undertake it. Thus, maintainability can be graphically presented by Fig. 3.2, where T represents the instant of time when the maintenance task has been completed. The restorability, which is an unknown value, is identical for all items under consideration, and therefore there is no need to assign it a numerical value.

Despite the fact that Fig. 3.2 is only an illustrative attempt to define the meaning of maintainability, it also suggests that the ability to restore functionability, by performing a specified maintenance task, could be numerically expressed by the shaded area. This means that maintainability is indirectly proportional to the area considered, and that the item with more desirable maintainability will cover a smaller area, and vice versa. It is necessary to stress that the size of the area depends mainly on decisions taken during the design phase. Decisions regarding the order of magnitude of the time required for restoration to functionability (5 minutes, 5 hours, or 2 days) can only be taken at a very early stage of the design process, through considerations related to the complexity of the maintenance task,

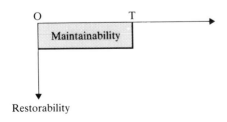

Figure 3.2 The concept of maintainability of an engineering item.

accessibility of the items, safety of the restoration, and the decisions related to the requirements for the maintenance support resources. Thus, maintainability can be quantitatively expressed through the length of time T for which the specified maintenance task is performed and the specified support resources used.

The question which immediately arises here concerns *the nature of T*. In other words, is T constant for each execution of the maintenance task or does it differ from trial to trial?

As an item exists only through the copies made, the maintenance task exists only through the physical execution of its consisting activities, on the copies. Thus, the answer to the above question depends on the times of each trial. In spite of the fact that each maintenance task consists of specified activities, performed in a specified sequence, the time needed for their execution might differ from trial to trial. To illustrate this point Table 3.1 gives the lengths of time in seconds taken for the replacement of a fog-light bulb by a group of second-year students from the School of Engineering, Exeter University.

In general, if the restoration time for several trials of a specific maintenance task is analysed it can be seen that one of them will be completed at the instant denoted by b_1, another at instant b_2 and say the nth will be executed by instance b_n, see Fig. 3.3. This illustration simply confirms what everyone familiar with the maintenance of engineering items already knows—the execution of each trial will be completed after a different interval of time. Thus, the length of time needed to complete each maintenance task is a specific characteristic of each trial.

The question which naturally arises is why different lengths of time are needed for the execution of the identical maintenance tasks. In order to

Table 3.1 Time taken to replace the bulb in a fog light

Restoration time	Number of results
<40	1
40–44	3
45–59	9
50–54	13
55–59	18
60–64	20
65–69	12
70–74	8
75–79	5
80–84	3
85–89	2
90	1

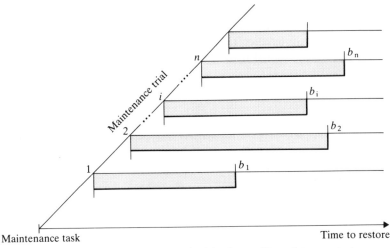

Figure 3.3 Maintenance pattern for several trials of a specific maintenance task.

provide the answer it is necessary to analyse all contributing factors. The following three groups are found to be the most influential:

- *Personal factors*—these represent the influence of the skill, motivation, experience, physical ability, self-discipline, responsibility, and other similar characteristics of the personnel involved.
- *Conditional factors*—these represent the influence of the operating environment and the consequences of the failure to the physical condition, geometry, and shape of the item under restoration.
- *Environmental factors*—these include temperature, humidity, lighting, vibration, time of the day, time of the year, wind, noise, and similar factors which affect maintenance personnel during the restoration.

Consequently, the nature of T for the maintenance task also depends on the variability of those parameters and the relationship between the influential factors, and can be expressed:

$$T = f \ (personal, \ conditional \ and \ environmental \ factors) \qquad (3.1)$$

Analysing the above expression it could be said that as a result of the large number of influential parameters in each group, in conjunction with their variability, it is impossible to find the rule which would deterministically describe the very complex relation denoted by f. The only way forward with maintainability analysis is to call upon *probability theory* as the tool to provide a probabilistic description of the relationship defined by Eq. (3.1).

In conclusion it can be said that it is impossible to give a deterministic answer regarding the instant of time when the transition from *SoFa* to *SoFu* will occur for any individual trial of a maintenance task being examined.

It is only possible to assign a certain probability that it will happen after a certain instant of maintenance time, or that a certain percentage of trials will or will not be completed by a specific instant in time.

FOUR

THE CONCEPT OF SUPPORTABILITY

- How long is the support task going to last?

The above question relates to the lower part of the functionability profile shown by Fig. 4.1. In the majority of cases the answer to this and similar questions cannot be found in glossy catalogues. The reasons for this are that support for a product during its operational life is almost completely ignored by designers, and that the communication between designers and users is almost non-existent.

Figure 4.1 Uncertain length of support time.

Today the situation is gradually changing, thanks to military and aerospace users who recognize the importance of information which relates to questions concerning operation and maintenance activities. In the absence of methods with which to assist designers and users obtain numerical answers to the above considerations, the need arose to form a new discipline and *supportability engineering* evolved. This can be defined as:

> A scientific discipline which studies the processes, activities and factors related to the support of a product with required resources for the execution of specified maintenance tasks, and works out methods for their quantification, assessment, prediction and improvement.

Nowadays supportability engineering is playing a leading role in the life cycle considerations of a product because it is recognized as making a considerable contribution towards the shape of the functionability profile and, as a consequence, the operational cost.

4.1 DEFINITION OF SUPPORTABILITY

In technical literature the complete definition for supportability is not clearly defined. Consequently, in this book the following definition is used:

> Supportability is the inherent characteristic of an item related to its ability to be supported by the required resources for the execution of the specified maintenance task.

To be applicable to everyday engineering practice, supportability, as defined above, has to be described in numerical terms. This means that the qualitative description has to be translated into quantitative measures.

In order to explain the physical meaning of supportability, let us establish the link between the maintenance process and the additional length of time during which the item is in *SoFa*. Thus, supportability can be graphically presented as shown by Fig. 4.2, where T represents the instant of time when the required support resources have been made available and the specified maintenance task can be performed. Despite this being only an illustrative attempt to define the meaning of supportability, it also suggests that the capability of an item to be supported during the execution of the maintenance task could be expressed, numerically, through the shaded area. This means that supportability is indirectly proportional to the area considered, and that the more supportable an item is the smaller the area, and vice versa. It is necessary to stress that the size of the area considered depends mainly on decisions taken during the design stage, being related to the complexity, size, quantity, and standardization of support resources.

Thus, supportability can be quantitatively expressed through the additional length of time T during which the support task is performed (for the item under consideration), and the maintenance activities that cannot

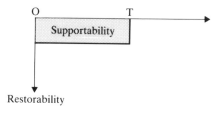

Figure 4.2 The concept of supportability of an engineering item.

be performed, due to lack of essential support resources (spares, material, personnel, facilities, software, equipment, tools, and similar).

The question that comes immediately to mind here concerns *the nature of T*. Is T constant for each execution of the specified support task or will it differ from trial to trial?

Since the item under consideration physically exists through the copies made, and the maintenance task exists only through physical execution of its consisting activities, the support task exists only through the repetitions of its trials. In spite of the fact that each support trial consists of several activities, related to the provision of support resources, the length of time required for their execution will differ from trial to trial. In order to clarify this point let us use a very simple example related to the additional lengths of time which several copies of a particular make and model spent in *SoFa*, due to delays caused by the lack of necessary support resources for clutch repairs. The data are shown in Table 4.1.

In general, if we analyse the additional length of time which several copies of the item under consideration spend in *SoFa* due to unavailability of maintenance support resources, it is possible to see that one of them will be supported by the instant denoted by c_1, another item by the instant c_2 and say the nth will be supported fully by the instant c_n, see Fig. 4.3. This illustration serves to confirm what everyone familiar with support of maintenance already knows, which is that the additional time spent in *SoFa*, due to the performance of the support task, differs from trial to trial. Thus, the additional time spent by each copy in *SoFa*, due to lack of supporting resources, is a specific characteristic of individual trials. The question which naturally follows is *why different lengths of time are needed for the execution of identical support tasks*.

To provide the answer it is necessary to analyse all the contributing factors, and experience has shown that the following groups are the most

Table 4.1 Waiting times for clutch repairs

Number of hours	Number of results
less than 1	18
1–3	17
4–6	14
7–10	7
11–14	4
14–18	2
19–24	4
24–36	1
37–48	0
48–72	2

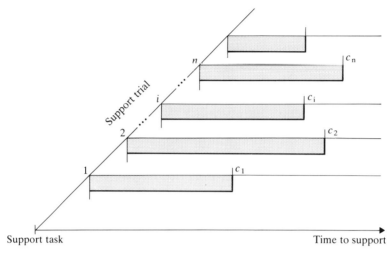

Figure 4.3 Length of time in *SoFa* for several copies of an engineering item.

influential:

- *Maintenance factors*—these are related to the management of the maintenance process, in particular its concept, policy and strategy.
- *Location factors*—the influence of the geographical location of items, communication systems, transport.
- *Investment factors*—these influence the provision of support resources (spares, tools, equipment, facilities).
- *Organizational factors*—these determine the flow of information and support elements.

Consequently, the nature of parameter T for the support task depends on the variability of all the above factors. The relationship between the influential factors and parameters T can be expressed as:

$$T = f \text{ (maintenance, location, investment organizational factors)} \quad (4.1)$$

Taking into account the analysis performed so far it could be concluded that T has an unpredictable nature, being the result of the variability and complexity of all influential factors to the restoration process, together with the provision of support resources. It is therefore reasonable to say that it is impossible to give a deterministic answer regarding the additional length of time which any specific item will spend in the state of failure. It is only possible to assign a probability that it will happen at a given instant of time, or that a certain percentage of trials will, or will not, be completed during a specific time interval.

FIVE

THE CONCEPT OF THE PROBABILITY SYSTEM

Probability theory plays a leading role in modern science in spite of the fact that it was initially developed as a tool which could be used for guessing the outcome of some games of chance. Probability theory is applicable to everyday life situations where the outcome of a repeated process, experiment, test, or trial is uncertain and a prediction has to be made.

In order to apply probability to everyday engineering practice it is necessary to learn the terminology, definitions and rules of probability theory. It is important to understand that this book is not intended to be a rigorous treatment of all relevant theorems and proofs. The intention is to provide an understanding of the main concepts behind probability theory and to show practical applications of the existing theorems and rules to definitions of reliability, maintainability and supportability measures in everyday engineering practice. As set theory provides the foundation to probability theory, it might be beneficial to read Appendices A_1 to A_4, before proceeding further, where the rudiments of set theory are given.

5.1 RELEVANT TERMS AND DEFINITIONS

In this chapter those elements essential to understanding the rudiments of elementary probability theory will be discussed and defined in a general manner, together with illustrative examples related to engineering practice. To facilitate the discussion some relevant terms and their definitions are introduced.

> **Experiment**—An experiment is a well-defined act or process that leads to a single well-defined outcome.

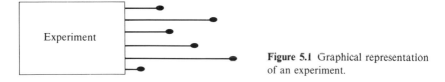

Figure 5.1 Graphical representation of an experiment.

This definition is generally accepted terminology in probability theory, to represent any process, trial, action or activity related to a real-life situation. Thus, every experiment must:

1. Be capable of being described, so that the observer knows when it occurs.
2. Have one and only one outcome, so that the set of all possible outcomes can be specified.

Figure 5.1 tries to illustrate this abstract concept of an experiment.

> **Elementary event**—An elementary event is every separate result of an experiment.

Making use of the definition for an experiment, it is possible to conclude that the total number of elementary events is equal to the total number of possible outcomes, since every experiment must have only one outcome.

> **Sample space**—The set of all possible distinct outcomes for an experiment is called the sample space for that experiment.

Most frequently in the literature the symbol S is used to stand for the *sample space*, and small letters, a, b, c, ... for elementary events that are possible outcomes of the experiment under consideration. The set S may contain either a finite or an infinite number of elementary events. Figure 5.2 is a graphical presentation of the sample space. Capital letters $A, B, C, ...$ are usually used for denoting events.

> **Event**—Each and every subset formed from the elementary events in S is an event.

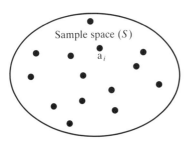

Figure 5.2 Graphical presentation of the sample space.

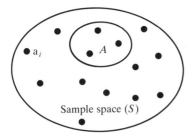

Figure 5.3 Graphical representation of a sample and its elements.

For example, if the experiment performed is measuring the speed of passing cars at a specific road junction, then the elementary event is the speed measured, whereas the sample space consists of all the different speeds one might possibly record. All speed events could be classified in, say, four different speed groups: A (less than 30 km/h), B (between 30 and 50 km/h), C (between 50 and 70 km/h) and D (above 70 km/h). If the measured speed is 35 km/h, then the event B is said to occur. An illustration of sample space, elementary events and events is given in Fig. 5.3.

Since *any* subset of the sample space is an event, then an event *must* occur on each and every trial of the experiment. As any event A is a subset of S, the operation of set theory is directly applicable to operations of events in order to define other events.

Suppose that we had a list containing every model and type of car registered in the United Kingdom. If we stand at a road junction at a specified time and observe the first car which passes, the elementary event is the actual model we see as the result of that observation, and the set S is the total set of all possible models. Suppose that, among all possible events, A is the set diesel, and B is Ford. If our observed model turns out to be a diesel, then event A occurs; if not, event A'. If it turns out to be a Ford, this is an occurrence of event B; if not, event B'. If the observation shows up as both diesel and Ford, $A \cap B$ occurs; if either diesel or Ford, then event $A \cup B$ occurs. If the observed model is diesel but not Ford, this is an occurrence of the event $A - B$. Only one thing is sure; we will observe a model registered in the United Kingdom, and the event S must occur.

Continuing with this example, suppose that the event C is engines under 1.5 l, event D is engines between 1.5 l and 2.9 l, and event E is engines over 3.0 l. The events C, D and E are mutually exclusive, since no two of them can occur at once. Also, they are exhaustive, since one of them must occur. That is, each model must fall into one and exactly one of the three classes or events. Thus, the events C, D, and E form a three-fold partition of the sample space S.

In the above example, the choice of the five events A, B, C, D and E was quite arbitrary, and would have been equally applicable to any other scheme for arranging elementary events into events.

5.2 ELEMENTARY THEORY OF PROBABILITY

The theory of probability is developed from axioms in the same way as algebra and geometry. In practice this means that its elements have been defined together with several axioms which govern their relations. All other rules and relations are derived from them. The full derivation of elementary rules and axioms can be found in Kolmogorov (1950).

5.2.1 Axioms of probability

In cases where the outcome of an experiment is uncertain it is necessary to assign some measure which will indicate the chances of occurrence of a particular event. Such a measure of events is called the *probability of the event* and symbolized by $P(.)$, and for event A is $P(A)$. The function which associates each event A in the sample space S with the probability measure $P(A)$, the probability of that event, is called the *probability function*. A graphical representation of the probability function is given in Fig. 5.4. Formally, the probability function is defined as:

A function which associates with each event A a real number $P(A)$, the probability of event A, such that the following axioms are true:

1. $P(A) > 0$ for every event A
2. $P(S) = 1$
3. The probability of the union of mutually exclusive events is the sum of their separate probabilities

$$P(A_1 \cup A_2 \cup \ldots \cup A_n) = P(A_1) + P(A_2) + \ldots + P(A_n) \tag{5.1}$$

In essence, this definition states that each event A is paired with a non-negative number, probability $P(A)$, and that the probability of the sure event S, or $P(S)$, is always 1. Furthermore, if A_1 and A_2 are any two

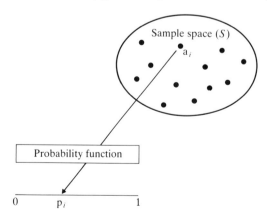

Figure 5.4 Graphical representation of a probability function.

mutually exclusive events in the sample space, the probability of their union $P(A_1 \cup A_2)$, is simply the sum of their two probabilities, $P(A_1) + P(A_2)$.

Events have probabilities, and in order to discuss probability it is necessary to discuss the events to which these probabilities are related. For example, if we speak of the probability that a car produced in the United Kingdom is a Ford, we are speaking of the probability number assigned to the event Ford in the sample space 'all cars produced in the UK'.

5.2.2 Rules of probability

The following elementary rules of probability are directly deduced from the original three axioms, utilizing set theory:

1. For any event A, the probability of the complementary event 'not A', written A', is one minus the probability of A

$$P(A') = 1 - P(A) \qquad (5.2)$$

2. The probability of any event must lie between zero and one inclusive

$$0 \le P(A) \le 1 \qquad (5.3)$$

3. The probability of an empty or impossible event is zero

$$P(\phi) = 0 \qquad (5.4)$$

4. If occurrence of an event A implies that an event B occurs, so that the event class A is a subset of event class B, then the probability of A is less than or equal to the probability of B

$$P(A) < P(B) \qquad (5.5)$$

5. In order to find the probability that A or B or both occur, the probability of A, the probability of B, and also the probability that both occur must be known, thus

$$P(A \cup B) = P(A) + P(B) - P(A \cap B) \qquad (5.6)$$

If A and B are *mutually exclusive* events, so that $P(A \cap B) = 0$, then

$$P(A \cup B) = P(A) + P(B) \qquad (5.7)$$

6. If n events form a partition of S, then their probabilities must add up to one:

$$P(A_1) + P(A_2) + \ldots + P(A_n) = \sum_{i=1}^{n} P(A_i) = 1 \qquad (5.8)$$

5.2.3 Joint events

• Any event that is an intersection of two or more events is a joint event.

There is nothing to restrict any given elementary event from the sample space from qualifying for two or more events, provided that those events are not mutually exclusive. Thus, given the event A and the event B, the joint event is $A \cap B$. The event $A \cap B$ occurs when any elementary event belonging to this set is actually observed.

Since a member of $A \cap B$ must be a member of set A, and also of set B, both A and B events occur when $A \cap B$ occurs. Furthermore, a member of $A \cap B$ is a member of $A \cup B$ and so whenever the joint event occurs, $A \cup B$ occurs as well. Provided that the elements of set S are all equally likely to occur, the probability of the joint event could be found in the following way

$$P(A \cap B) = \frac{\text{number of elementary events in } A \cap B}{\text{total number of elementary events}}$$

5.2.4 Conditional probability

• If A and B are events in a sample space which consists of a finite number of elementary events the conditional probability of B given A, denoted by $P(B|A)$, is defined as

$$P(B|A) = \frac{P(A \cap B)}{P(A)}, \qquad P(A) > 0 \tag{5.9}$$

The conditional probability symbol, $P(B|A)$, is read as 'the probability of B given A'. It is necessary to satisfy the condition that $P(A) > 0$, because it does not make sense to consider the probability of B given A if event A is impossible. For any two events A and B, there are two conditional probabilities that may be calculated

$$P(B|A) = \frac{P(A \cap B)}{P(A)} \qquad \text{and} \qquad P(A|B) = \frac{P(A \cap B)}{P(B)}$$

One of the most important and frequently used expressions related to conditional probability is Baye's theorem whose definition can be presented in the following way

If $(A_1, A_2, \ldots, A_N)$ presents a set of N mutually exclusive events, and if B is a subset of $A_1 \cup A_2 \cup \ldots \cup A_N$, so that $P(B) \subset P(A_1) + P(A_2) + \ldots + P(A_N)$, then

$$P(A_i|B) = \frac{P(B|A_i)P(A_i)}{P(B|A_1)P(A_1) + \ldots + P(B|A_i)P(A_i) + \ldots + P(B|A_N)P(A_N)} \tag{5.10}$$

The graphical presentation of this theorem is given in Fig. 5.5.

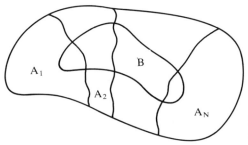

Figure 5.5 Graphical presentation of Baye's theorem.

Example 5.1 The same compressor is manufactured at three plants. Plant 1 makes 70 per cent of the requirement and plant 2 makes 20 per cent. From plant 1, 90 per cent of compressors meet a particular standard, from plant 2 only 80 per cent, whereas from plant 3 the percentage of units produced up to standard is 95. Determine:

(a) How many compressors will be up to standard out of every 100 purchased by customers?

(b) Given that the purchased compressor is up to standard, what is the probability that it was made in plant 2?

SOLUTION

(a) Applying the concept of conditional probability, let us use the following notation: A is the event that the compressor is up to standard, B_1 is the event that the item is made in plant 1, B_2 and B_3 are the events that the compressor is made in plants 2 and 3.

Consequently: $P(A|B_1) = 0.9,$ $P(A|B_2) = 0.8,$ $P(A|B_3) = 0.95$
$P(B_1) = 0.7,$ $P(B_2) = 0.2,$ $P(B_3) = 0.1$

$P(A) = 0.9 \times 0.7 + 0.8 \times 0.2 + 0.95 \times 0.1 = 0.63 + 0.16 + 0.095 = 0.885$, and out of every 100 items purchased by the customer, $100 \times 0.885 = 88$ will be up to standard.

(b) According to part (a), the probability that the purchased compressor is up to standard and comes from plant 2, $P(A \cap B_2) = 0.16$, and the probability that it is up to standard, $P(A) = 0.88$. Thus

$$P(B_2|A) = \frac{P(A \cap B_2)}{P(A)} = \frac{0.16}{0.88} = 0.18$$

5.2.5 Probability and experimental data

The connection between probability and long-run relative frequency forms a tie between that basically undefined notion the 'probability of an event', and something which can actually be observed, a relative frequency of occurrence. A statement of probability tells us what to *expect* about the relative frequency of occurrence given that enough observations are made. In the long-run, the relative frequency of occurrence of an event, say A, should approach the probability of this event, if independent trials are made at random over an indefinitely long sequence.

This principle was first formulated and proved by James Bernoulli in the early eighteenth century, and is now well-known as Bernoulli's theorem:

If the probability of occurrence of an event A is p, and if n trials are made, independently and under the same conditions, then the probability that the relative frequency of occurrence of A, defined as $f(A) = N(A)/n$ differs from p by any amount, however small, approaches zero as the number of trials grows indefinitely large.

The theorem may also be represented by the following mathematical expression:

$$P(|N(A)/n) - p| > s) \to 0, \quad \text{as} \quad n \to \infty \tag{5.11}$$

where s is some arbitrarily small positive number.

This does not mean that the proportion of $N(A)/n$ occurrences among any n trials *must* be p; the proportion actually observed might be any number between 0 and 1. Nevertheless, given more and more trials, the relative frequency of $f(A)$ occurrences may be expected to come closer and closer to p.

Although it is true that the relative frequency of occurrence of any event is exactly equal to the probability of occurrence of any event only for an infinite number of independent trials, this point must not be overstressed. Even with a relatively small number of trials there is very good reason to expect the observed relative frequency to be quite close to the probability because the rate of convergence of the two is very rapid.

Example 5.2 A firm manufacturing a nominal 50 kW motor car engine, produced in one year 10 000 such engines. The performance of each engine produced was measured and classified according to the achieved output power. The number of engines falling into the various kW classifications to one decimal place of measurement were found to be as shown in Table 5.1. If the same standard of production is achieved in the future, what is the probability of having an engine:

(a) Greater than or equal to 50 kW?
(b) Whose power output lies between 49 and 50.9 kW?
(c) Which is either less than 48 kW or more than 51.9 kW?

SOLUTION Each of the power ranges can be considered as an event in the sample space which consists of six events, say A, B, C, D, E and F. The probability of each event is estimated from the number of engines falling in a particular range. Hence, a table of

Table 5.1 Engine power in kW

Power achieved	47.0 47.9	48.0 48.9	49.0 49.9	50.0 50.9	51.0 51.9	52.0 52.9
No. of engines	84	761	6852	2148	139	16
Range	1	2	3	4	5	6

probabilities can be constructed:

Range	1	2	3	4	5	6
Event	A	B	C	D	E	F
Probability	0.0084	0.0761	0.6852	0.2148	0.0139	0.0016

Each probability represents a mutually exclusive event since a particular outcome can only fall into one of the events. Any combination of the probabilities, therefore, will follow the summation rule.

(a) The probability of having an engine greater than or equal to 50 kW is given by the summation of the probabilities for events D, E, and F

$$P = P(D) + P(E) + P(F) = 0.2148 + 0.0139 + 0.0016 = 0.2303$$

(b) The probability of having an engine whose power output lies between 48.9 and 51 kW is given by the summation of the probabilities for events C and D

$$P = P(C) + P(D) = 0.6852 + 0.2148 = 0.900$$

(c) The probability of having an engine which is either less than 48 kW or more than 51.9 kW is obtained from the sum of the probabilities for events A and F

$$P = P(A) + P(F) = 0.0085 + 0.0016 = 0.00101$$

Example 5.3 A factory producing automobile tyres has an average output of 1.5 million tyres per year. Over a 10-year period it has been found that 9000 tyres have been rejected during the final test and inspection as substandard.

(a) What is the probability that any tyre being inspected will be rejected?
(b) What is the probability that any tyre sold will be substandard if the inspection department only detects 97 per cent of all substandard tyres?

SOLUTION
(a) $n = 10 \times 1500000 = 15000000$, $\qquad N(A) = 9000$

$$P(\text{tyre rejected}) = \frac{9000}{15\,000\,000} = 0.0006$$

(b) If the 9000 tyres represent 97 per cent of all the substandard tyres produced, then the number of rejects, $N(B)$, which passed inspection was

$$N(B) = 9000 \times \frac{3}{97} = 278$$

the estimated probability that a customer will receive a reject tyre is

$$P(\text{customer receiving a reject tyre}) = \frac{N(B)}{n} = \frac{278}{15\,000\,000} = 0.0000185$$

Example 5.4 For a satellite about to be launched an estimate is given, on the basis of past performance, that its chance of success will be 0.68. If a total of 25 satellites of the same type had been previously launched under the same conditions, what is the number of unsuccessful missions that have already occurred?

SOLUTION

$P(\text{success}) = 0.68,$

$P(\text{unsuccessful}) = 1 - P(\text{success}) = 1 - 0.68 = 0.32$

$P(\text{unsuccessful}) = \dfrac{N(\text{unsuccessful})}{(\text{total number of trials})}, \text{ thus}$

$N(\text{unsuccessful}) = P(\text{unsuccessful}) \times (\text{total number of trials}) = 0.32 \times 25 = 8$

5.3 PROBABILITY DISTRIBUTION

- Any statement of a probability function having a set of mutually exclusive and exhaustive events for its domain is a probability distribution.

Consider the set of events $A_1, A_2, \ldots, A_n$, and suppose that they form a partition of the sample space S. That is, they are mutually exclusive and exhaustive. The corresponding set of probabilities, $P(A_1), P(A_2), \ldots, P(A_n)$, is a probability distribution. An illustrative presentation of the concept of probability distribution is shown in Fig. 5.6.

As a simple example of a probability distribution, imagine a sample space of all produced Ford cars. A car selected at random is classified as a saloon or coupe or estate. The probability distribution might be:

Event	Saloon	Coupe	Estate	Total
P	0.60	0.31	0.09	1.00

All events other than those listed have probabilities of zero.

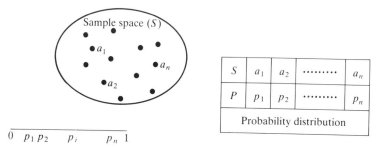

Figure 5.6 Graphical representation of probability distribution.

5.4 THE CONCEPT OF THE RANDOM VARIABLE

- A function that assigns a number (usually a real number) to each sample point in the sample space S is a random variable.

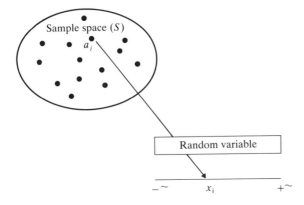

Figure 5.7 Graphical representation of the random variable.

Outcomes of experiments may be expressed in numerical and non-numerical terms. In order to compare and analyse them it is much more convenient to deal with numerical terms. So, from the point of view of practicality, it is necessary to assign a numerical value to each possible elementary event in a sample space S. Even if the elementary events themselves are already expressed in terms of numbers, it is possible to reassign a single number to each elementary event. Thus, each elementary event in S can be associated with one, and only one, real number. The function which achieves this is known as the *random variable*.

Suppose that the symbol X is used to stand for any particular number assigned to any given elementary event. Thus, X is a variable since there will be some set of elementary events assigned to this value. In other words, a random variable is a real-valued function defined in a sample space. Usually it is denoted with capital letters, such as X, Y and Z, whereas small letters, such as x, y, z, a, b, c, and so on, are used to denote particular values of random variables, see Fig. 5.7.

If X is a random variable and r is a fixed real number, it is possible to define the event A to be the subset of S consisting of all sample points a to which the random variable X assigns the number r, $A = (a : X(a) = r)$. On the other hand, the event A has a probability $p = P(A)$. The symbol p can be interpreted, generally, as the probability that the random variable X takes on the value r, $p = P(X = r)$. Thus, the symbol $P(X = r)$ represents the probability function of a random variable.

Therefore, by means of the random variable it is possible to assign probabilities to real numbers, although the original probabilities were only defined for events of the set S, as shown in Fig. 5.8.

Suppose that X symbolizes the speed of a motor vehicle on a particular road junction, measured to the nearest km/h. Here, there is some probability that $X < 60$, or $P(X < 60)$, since there is presumably a set of vehicles with speed less than or equal to 60 km/h. Furthermore, there is some probability $P(70 < X < 90)$, since there is a set of vehicles having speeds between

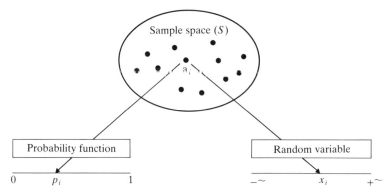

Figure 5.8 Relationship between a probability function and the random variable.

70 and 90 km/h inclusive. There are also some probabilities $P(X < 1)$ and $P(X < 300)$. These probabilities are almost certainly zero and one, respectively, but the point is that they *do exist*. Thus, X symbolizes the numerical value (speed in km/h) assigned to a passing vehicle (an elementary event).

In conclusion, it could be said that the symbol X represents any of many different values which a random variable under consideration can take. Consequently, for any arbitrary number, say a, there is a probability that the value of X is less than or equal to that value, $P(X < a)$. This expression is called the *distribution function of the random variable X*.

5.5 TYPES OF RANDOM VARIABLES

According to the values which they can assume, random variables can be classified as discrete or continuous. The main characteristics, similarities and differences for both types will be briefly described below.

5.5.1 Discrete random variables

- If the random variable X can assume only a particular finite or countably infinite set of values, it is said to be a discrete random variable.

There are very many situations where the random variable X can assume only a particular *finite* or *countably infinite* set of values; that is, the possible values of X are finite in number or they are infinite in number but can be put in a one-to-one correspondence with the positive integers. For example, suppose that a simple experiment consists of selecting one British family for observation and noting the number of cars in that family; an elementary event is one particular British family, and associated with each family is a number X. Here, X can assume only certain values between, say, 0 and 30, and these values must be integers. The numbers 3.68, -37 or 0.21 simply cannot be associated appropriately with any elementary event in the sample space. The random variable X can assume only a *finite* set of values.

As another example, suppose that a simple experiment consists of starting the engine of a particular car repeatedly until it fails to start. Let X be the number of starts it takes to get the first failure. The first failure could appear on the first trial, in which case X would be 1; it might not appear until the one-hundredth trial, in which case X would be 100; in general, it might not appear until the kth trial, where k is any whole number. Thus, we have a *countably infinite* set of values for X.

5.5.2 Continuous random variables

- If the random variable X can assume any value from a finite or an infinite set of values, it is said to be a continuous random variable.

Let us consider an experiment which consists of recording the temperature of the cooling liquid of an engine in the area of the thermostat at a given time. Suppose that we can measure the temperature exactly, which means that our measuring device allows us to record the temperature to any number of decimal points. If X is the temperature reading, it is not possible for us to specify a finite or countably infinite set of values. For example, if one of the finite set of values is 75.965, we can determine values 75.9651, 75.9652, and so on, which are also possible values of X. What is being demonstrated here is that the possible values of X consist of the set of real numbers, a set which contains an infinite (and uncountable) number of values.

It is very difficult to find real examples of continuous random variables. This concept is really an idealization that is never quite true, but which has enormous mathematical utility.

5.6 PROBABILITY DISTRIBUTION OF A RANDOM VARIABLE

Taking into account the concept of the probability distribution and the concept of the random variable, it could be said that the probability distribution of the random variable is a set of pairs, $\{r_i, P(X = r_i)\}, i = 1, n$ as shown in Fig. 5.9. The easiest way to present this set is to make a list of all its members. If the number of possible values is small, it is easy to specify a probability distribution. On the other hand, if there are a large number of possible values a listing may become very difficult. In the extreme case where we have an infinite number of possible values (for example, all real

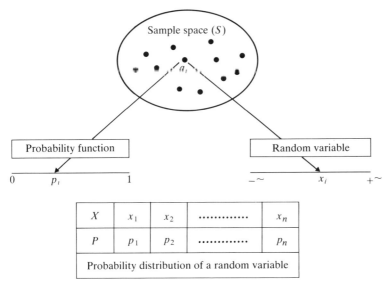

Figure 5.9 Probability distribution of a random variable.

numbers between zero and one) it is clearly impossible to make a listing. Fortunately, there are other methods which could be used for specifying a probability distribution of a random variable:

(a) *Functional method*—where specific mathematical functions exist, from which the probability of any value or interval of values can be calculated.
(b) *Parametric method*—where the entire distribution is represented through one or more parameters, known as summary measures.

Both methods for expressing probability distributions of random variables will be examined below.

5.7 FUNCTIONAL METHOD

By definition a function is a relation where each member of the domain is paired with one member of the range. In this particular case the relation between numerical values which random variables can take, and their probabilities, will be considered. The most frequently used functions for the description of probability distribution of a random variable are:

(a) probability mass function,
(b) probability density function,
(c) cumulative distribution function.

Each of these will be analysed and defined in the remainder of this chapter.

5.7.1 Probability mass function

This function is related to a discrete random variable and it represents the probability that the random variable, X, will take one specific value x_i, $p_i = P(X = x_i)$. Thus, a probability mass function, which is usually denoted as $PMF(.)$, places a mass of probability p_i at the point x_i on the X-axis.

Given that a discrete random variable takes on only n different values, say $a_1, a_2, \ldots, a_n$, the corresponding $PMF(.)$ must satisfy the following two conditions:

1. $P(X = a_i) \geq 0,$ for $i = 1, 2, \ldots, n$

2. $\sum\limits_{i=1}^{n} P(X = a_i) = 1.$ (5.12)

In practice this means that the probability of each value that X can take must be non-negative and the sum of the probabilities over all different values must be 1. Thus, a probability distribution can be represented by the set of pairs of values (a_i, p_i), where $i = 1, 2, \ldots, n$, as shown in Fig. 5.10. The advantage of such a graph over a listing is the ease of comprehension and a better provision of a notion for the nature of the probability distribution.

Figure 5.10 Probability mass function.

5.7.2 Probability density function

In the previous section, discrete random variables were discussed in terms of probabilities $P(X = x)$, the probability that the random variables take on an *exact* value. However, consider the example of an infinite set for a specific type of car where the volume of the fuel in the fuel tank is measured with any degree of accuracy. What is the probability that a car selected at random will have *exactly* 16 litres of fuel? This could be considered as an event which is defined by the interval of values between, say 15.5 and 16.5, or 15.75 and 16.25, or any other interval $16 \pm 0.1 \times i$, where i is not exactly zero. Since the smaller the interval, the smaller the probability, the probability of exactly 16 litres is, in effect, zero.

> In general, for continuous random variables, the occurrence of any exact value of X may be regarded as having zero probability.

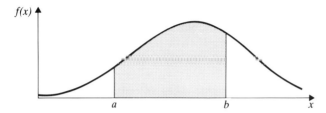

Figure 5.11 Probability density function for a hypothetical distribution.

As a consequence, the probabilities of a continuous random variable can be discussed only for *intervals* of X values. Thus, instead of the probability that X takes on a specific value, say a, we deal with the so-called *probability density* of X at a, symbolized by $f(a)$. In general, the probability distribution of a continuous random variable can be represented by its *probability density function, PDF*, which is defined in the following way

$$P(a \leq X \leq b) = \int_a^b f(x)dx \qquad (5.13)$$

A fully defined probability density function must satisfy the following two requirements

1. $f(x) \geq 0,$ for all x.
2. $\int_{-\infty}^{+\infty} f(x)dx = 1$ $\qquad (5.14)$

The PDF is always represented as a smooth curve drawn above the horizontal axis, which represents the possible values of the random variable X. A curve for a hypothetical distribution is shown in Fig. 5.11 where the two points a and b on the horizontal axis represent limits which define an interval. The shaded portion between a and b represents the probability that X takes on a value between the limits a and b.

5.7.3 Cumulative distribution function

The probability that a random variable X takes on a value at or below a given number, a, is often written

$$F(a) = P(X \leq a) \qquad (5.15)$$

The symbol $F(a)$ denotes the particular probability for the interval $X \leq a$; the general symbol $F(x)$ is sometimes used to represent the function which relates the various values of X to the corresponding cumulative probabilities. This function is called the *cumulative distribution function, CDF*, and it must satisfy certain mathematical properties, the most important of which are

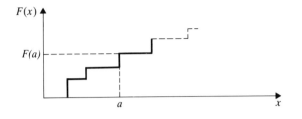

Figure 5.12 Cumulative distribution function for a discrete variable.

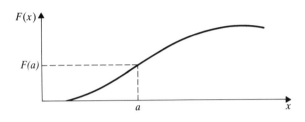

Figure 5.13 Cumulative distribution function for a continuous variable.

1. $0 \le F(x) \le 1$
2. If $a < b$, $\quad F(a) \le F(b)$
3. $F(\infty) = 1$ $\quad$ and $\quad F(-\infty) = 0$

The symbol $F(x)$ can be used to represent the cumulative probability that X is less than or equal to x. It is defined as

$$F(a) = \sum_{i=1}^{a} P(X = x_i) \qquad (5.16)$$

for discrete random variables, whereas in the case of continuous random variables it will take the following form

$$F(a) = \int_{-\infty}^{a} f(x)dx \qquad (5.17)$$

Hypothetical cumulative distribution functions for both types of random variables are given in Figs. 5.12 and 5.13.

5.8 PARAMETRIC METHOD

In some situations it is easier and even more efficient to look only at certain characteristics of distributions than to attempt to specify the distribution as a whole. Such characteristics summarize and numerically describe certain features for the entire distribution. Two general groups of such characteristics, applicable to any type of distribution are:

(a) *Measures of central tendency* (or location) which indicate the typical or the average value of the random variable.
(b) *Measures of dispersion* (or variability) which show the spread of the difference among the possible values of the random variable

In many cases, it is possible to adequately describe a probability distribution with a few measures of this kind. It should be remembered, however, that these measures serve only to summarize some important features of the probability distribution. In general, they do not completely describe the entire distribution.

One of the most common and useful summary measures of a probability distribution is that known as the *expectation* of a random variable, $E(X)$. It is a single number which indicates a location for the distribution as a whole. The concept of expectation plays an important role not only as a useful summary measure, but also as a central concept within the theory of probability and statistics.

If a random variable, say X, is discrete, then its expectation is defined as

$$E(X) = \sum_x x \times P(X = x) \tag{5.18}$$

where the sum is taken for all values that the variable X can assume. If the random variable is continuous, the expectation is defined as

$$E(X) = \int_{-\infty}^{+\infty} x \times f(x)dx \tag{5.19}$$

The similarity between discrete and continuous variables is clarified, because $E(X)$ in the former case is a sum of products of values of X and their probabilities, whereas in the latter case $E(X)$ is an integral (which is, after all, a sum in the limiting case) of products of values of X and their probability densities. Furthermore, it is possible to establish the relationship between the expectation of the random variable and its cumulative distribution function for both types of random variable. The practical benefit of this will be clearly demonstrated in future chapters.

If the random variable, X, is discrete, then its expectation, $E(X)$, is defined as

$$E(X) = \sum_x 1 - F(x) \tag{5.20}$$

where the sum is taken over all values that X can assume. For a continuous random variable the expectation is defined as

$$E(X) = \int_{-\infty}^{+\infty} [1 - F(x)]dx \tag{5.21}$$

5.8.1 Measures of central tendency

The most frequently used measures are:

(a) the mean, M,
(b) the median, Md,
(c) the mode, Mo.

A brief description of them follows.

- *The mean* of the distribution is simply the expectation of the random variable under consideration. Thus, for the random variable, X, the mean value is defined as

$$M = E(X) \tag{5.22}$$

- *The median* is defined as the value of X which is midway (in terms of probability) between the smallest possible value and the largest possible value. The median is the point which divides the total area under the *PDF* into two equal parts. In other words, the probability that X is less than the median is $1/2$, and the probability that X is greater than the median is also $1/2$. Thus, if $P(X \leq a) \geq 0.50$ and $P(X \geq a) \geq 0.50$ then a is the *median* of the distribution of X. In the continuous case, this can be expressed

$$\int_{-\infty}^{a} f(x)dx = \int_{a}^{+\infty} f(x)dx = 0.50 \tag{5.23}$$

and the median is a unique since $f(x)$ is a continuous function.
- *The mode* is defined as the value of X at which the *PMF* or *PDF* of X reaches its highest point. If a graph of the *PMF* (or *PDF*), or a listing of possible values of X along with their probabilities is available, determination of the mode is quite simple. However, if the distribution is expressed in functional form, determination of the mode may not be quite so easy.

A central tendency parameter, whether it is mode, median, mean, or any other measure, summarizes only a certain aspect of a distribution. It is easy to find two distributions which have the same mean but which are not at all similar in any other respect.

5.8.2 Measures of dispersion

The mean is a good indication of the location of a random variable, but *no single value need be exactly like the mean*. A deviation from the mean, D, expresses the measure of error made by using the mean as a particular value

$$D = X - M$$

The deviation can be taken from other measures of central tendency such as the median or mode.

It is quite obvious that the larger such deviations are from a measure of central tendency, the more the individual values differ from each other, and the more apparent the spread, within the distribution, becomes. Consequently, it is necessary to find a measure which will reflect the spread, or variability, of individual values.

The expectation of the deviation about the mean as a measure of variability, $E(X - M)$, will not work because the expected deviation from the mean must be zero for obvious reasons. The solution is to find the *square* of each deviation from the mean, and then to find the expectation of the squared deviation. This characteristic is known as the *variance of the distribution, V*, thus

$$V(X) = E(X - M)^2 = \sum (x - M)^2 \times P(x) \quad \text{if } X \text{ is discrete} \quad (5.24)$$

$$V(X) = E(X - M)^2 = \int_{-\infty}^{+\infty} (x - M)^2 \times f(x)dx \quad \text{if } X \text{ is continuous}$$

$$(5.25)$$

The variance presents the degree of the dispersion, because the more the values differ from each other, and the mean, the larger the variance becomes.

Regarding the establishment of the relationship between measures of dispersion and the cumulative distribution function, the situation is identical to the case for the expectation. Thus, the variance can be defined as

$$V(X) = E(X - M)^2 = 2 \times \sum_x x \times [1 - F(x)] - E(X) \quad \text{if } X \text{ is discrete}$$

$$(5.26)$$

$$V(X) = E(X - M)^2 = 2 \times \int_{-\infty}^{+\infty} x \times [1 - F(x)]dx - E(X) \quad \text{if } X \text{ is continuous}$$

$$(5.27)$$

The positive square root of the variance for a distribution is called the *standard deviation, SD*. This is a measure of variability which is defined by the following expression

$$SD = \sqrt{V(X)} \quad (5.28)$$

Probability distributions can be analysed in greater depth by introducing other summary measures, known as *moments*. Very simply these are expectations of different powers of the random variable. More information about them can be found in texts on probability.

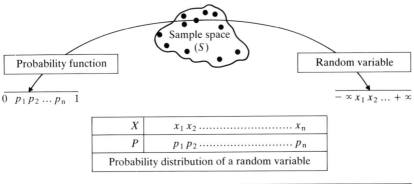

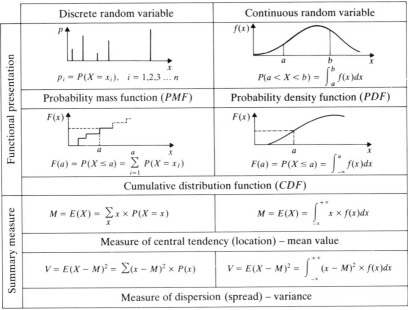

Figure 5.14 Probability system.

5.8.3 Variability

The standard deviation is a measure which shows how closely the values of random variables are concentrated around the mean. Sometimes it is difficult when using only a knowledge of the standard deviation, to decide whether the dispersion is considerably large or small, because this will depend on the mean value. In this case the parameter known as coefficient of variation, CV_X, defined as

$$CV_X = \frac{SD}{M} \tag{5.29}$$

is very useful because it gives better information regarding the dispersion.

5.9 GRAPHICAL PRESENTATION OF THE PROBABILITY SYSTEM

In order to clarify the concept of the probability system presented above, its graphical representation is given in Fig. 5.14 on page 43 as a flow diagram, for both types of random variable, discrete and continuous.

In conclusion it could be said that the probability system is wholly abstract and axiomatic. Consequently, every fully defined probability problem has a unique solution.

THEORETICAL PROBABILITY
DISTRIBUTIONS

In probability theory there are several rules which define the functional relationship between the possible values of random variable X and their probabilities, $P(X)$. Rules which have been developed by mathematicians will be analysed here. As they are purely theoretical, i.e. they do not exist in reality, they are called *theoretical probability distributions*. Instead of analysing the ways in which these rules have been derived the analysis in this chapter concentrates on their properties.

There are two types of theoretical probability distributions which correspond to the two types of random variable, namely:

1. Discrete theoretical probability distributions.
2. Continuous theoretical probability distributions.

It is necessary to emphasize that all theoretical distributions represent the family of distributions defined by a common rule through unspecified constants known as the *parameters of distribution*. The particular member of the family is defined by fixing numerical values for the parameters which define the distribution. The probability distributions most frequently used in reliability, maintainability and supportability engineering are examined in this chapter.

6.1 DISCRETE PROBABILITY DISTRIBUTIONS

Among the family of theoretical probability distributions which are related to discrete random variables, the binomial distribution and the Poisson distribution are relevant to the objectives set by this book. A brief description of each now follows.

6.1.1 The binomial distribution

- The theoretical probability distribution which pairs the number of successes in n trials with its probability is called the binomial distribution.

This probability distribution is related to experiments which consist of a series of independent trials, each of which can result in only one of two outcomes: success or failure. These names are used only to tell the events apart. By convention the symbol p stands for the probability of a success, q for the probability of failure $(p + q = 1)$.

The number of successes in n trials is a discrete random variable which can take on only the whole values from 0 through n. The formal rule for the probability mass function of the discrete random variable X is

$$PMF(x) = P(X = x) = \binom{n}{x} p^x q^{n-x}, \qquad 0 < x < n \qquad (6.1)$$

where

$$\binom{n}{x} p^x q^{n-x} = \frac{n!}{x!(n-x)!} p^x q^{n-x} \qquad (6.2)$$

The binomial distribution expressed in cumulative form, representing the probability that X falls at or below a certain value a, is defined by the equation

$$P(X \leq a) = \sum_{i=0}^{a} P(X = x_i) = \sum_{i=0}^{a} \binom{n}{i} p^i \times q^{n-i} \qquad (6.3)$$

As an illustration of the binomial distribution, the PMF and CDF are shown in Fig. 6.1 with parameters $n = 10$ and $p = 0.3$.

According to the expression for the expectation (see Eq. (5.18)), the mean of the distribution is the sum of the means of the individual trials; in the case of the binomial distribution it is p summed n times

$$E(X) = np \qquad (6.4)$$

Similarly, because of the independence of trials, the variance of the binomial distribution is the sum of the variances of the individual trials, or $p(1 - p)$ summed n times

$$V(X) = np(1 - p) = npq \qquad (6.5)$$

Consequently, the standard deviation is equal to $\sqrt{npq}$.

Although the mathematical rule for the binomial distribution is the same regardless of the particular values which parameters n and p take, the shape

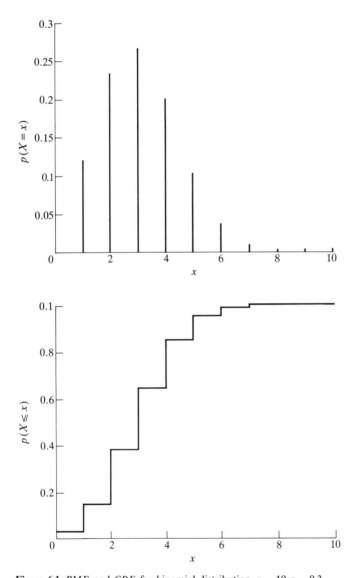

Figure 6.1 *PMF* and *CDF* for binomial distribution, $n = 10, p = 0.3$.

of the probability mass function and the cumulative distribution function will depend upon them. The *PMF* of the binomial distribution is symmetric if $p = 0.5$, positively skewed if $p < 0.5$, and negatively skewed if $p > 0.5$. Thus, the binomial distribution represents the family of distributions defined by the same rule, where a particular member of the family is defined by fixing numerical values for n and p.

Summary

$$PMF(x) = \binom{n}{x} p^x q^{n-x}$$

$$F(a) = \sum_{i=0}^{a} \binom{n}{i} p^i \times q^{n-i}$$

$$E(X) = np$$

$$V(X) = npq$$

Example 6.1 Consider a large fleet of cars, where 80 per cent of them have an engine which can only run on leaded petrol and 20 per cent of them can use unleaded petrol. What is the probability that an engineer checking three cars will find one with an unleaded petrol engine?

SOLUTION This may be regarded as a Bernoulli process with unleaded being a success and leaded a failure, with corresponding probabilities $p = 0.20$ and $q = 0.80$. The required probability can be determined by making use of Eq. (6.1), where $n = 3$, $x = 1$, thus

$$F(1) = P(X{=}1) = \binom{3}{1}(0.20)^1(0.80)^2 = 0.384$$

Example 6.2 According to past information the probability of producing a defective component is 0.05. Maintaining the same production process for six consecutive trials determine:

(a) The probability of no defectives;
(b) The probability of two defectives;
(c) The mean and the standard deviation of the number of defectives.

SOLUTION On the assumption that the probability of occurrence of a defective remains constant from trial to trial, the problem can be viewed as a series of six Bernoulli trials and the required answers can be obtained from the binomial distribution with $n=6$, $p=0.05$ and $q=0.95$. Thus

(a) $$P(X = 0) = \frac{n!}{x!(n-x)!}p^x q^{n-x} =$$

$$\frac{6!}{0!(6-0)!}\ 0.05^0 \times 0.95^6 = 0.7351$$

(b) $$P(X = 2) = \frac{6!}{2!(6-2)!}0.05^2 \times 0.95^4 = 0.0305$$

(c) Mean number of defectives $M(X) = np = 6 \times 0.05 = 0.30$
Standard deviation $SD(X) = \sqrt{npq} = \sqrt{6 \times 0.05 \times 0.95} - 0.5338$

Example 6.3 A product is claimed to be 90 per cent free of defects. What is the expected value and standard deviation of the number of defects in a sample of four?

SOLUTION In this example $n = 4$, $p = 0.10$ and $q = 1 - 0.10 = 0.9$

$$E(X) = 4 \times 0.10 = 0.4 \text{ and } SD(X) = \sqrt{4 \times 0.10 \times 0.90} = 0.6$$

Table 6.1 Solution to Example 6.3

Defects	Individual probability	$E(X)$	$E(X^2)$
0	0.6561	—	-
1	0.2916	0.2916	0.2916
2	0.0486	0.0972	0.1944
3	0.0036	0.0108	0.0324
4	0.0001	0.0004	0.0016
	1.0000	0.4000	0.5200

The same results can be achieved in a more tedious way by the direct application of Eqs (5.18) for $E(X)$ and (5.24) for $V(X)$, as shown in Table 6.1. Therefore, the expected number of defects in the sample of four is $E(X) = 0.4$ with a standard deviation $SD(X) = \sqrt{0.52 - 0.4^2} = 0.6$.

Example 6.4 It is known that screws produced by a certain company will be defective with a probability 0.01 independently of each other. The company sells the screws in packages of 12 and offers a money-back guarantee that at most one of them is defective. What proportion of packages sold must the company replace?

SOLUTION If X is the number of defective screws in a package, then X is a binomial random variable with parameters (12, 0.01). Hence the probability that a package will have to be replaced is

$$1 - P(X = 0) - P(X = 1) = 1 - \binom{12}{0} \times 0.01^0 \times 0.99^{12} - \binom{12}{1} \times 0.01^1 \times 0.99^{11} = 0.00426$$

Hence only less than one per cent of the packages will have to be replaced.

6.1.2 The Poisson distribution

• The theoretical probability distribution which pairs the number of occurrences of an event in a given time period duration with its probability is called the Poisson distribution.

There are experiments where it is not possible to observe a finite sequence of trials. Instead, observations take place over a continuum, such as time. For example, if the number of cars arriving at a specific junction in a given period of time is observed, say for one minute, it is difficult to think of this situation in terms of finite trials. If the number of binomial trials n is made larger and larger and p smaller and smaller, in such a way that np remains constant, then the probability distribution of the number of occurrences of the random variable approaches the Poisson distribution. The probability mass function in the case of the Poisson distribution for random variable X can be expressed

$$P(X = x \mid \lambda) = \frac{\exp(-\lambda)\lambda^x}{x!} \quad \text{where} \quad x = 0, 1, 2,\ldots$$

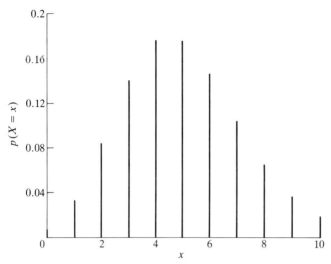

Figure 6.2 The *PMF* of the Poisson distribution with $\lambda = 5$.

Here, λ is the *intensity* of the process and represents the expected number of occurrences in a time period of length t. Figure 6.2 shows the *PMF* of the Poisson distribution with $\lambda = 5$

The cumulative distribution function for the Poisson distribution is

$$F(x) = P(X \le x) = \sum_{i=0}^{x} \frac{\exp(-\lambda)\lambda^i}{i!} \tag{6.6}$$

The *CDF* of the Poisson distribution with $\lambda = 5$ is presented in Fig. 6.3.

According to the definition of expectation given in Eq. (5.18)

$$E(X) = \sum_{x} xP(X = x) = \sum_{x=0} x\frac{\exp(-\lambda)\lambda^x}{x!}$$

Applying some simple mathematical transformations it can be proved that

$$E(X) = \lambda \tag{6.7}$$

which means that the expected number of occurrences in a period of time t is equal to np, which is equal to λ.

The variance of the Poisson distribution is equal to the mean

$$V(X) = \lambda \tag{6.8}$$

Thus, the Poisson distribution is a single parameter distribution because it is completely defined by the parameter λ. In general, the Poisson distribution is positively skewed, although it is nearly symmetrical as λ becomes larger.

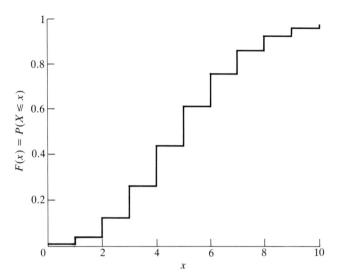

Figure 6.3 The *CDF* of the Poisson distribution with $\lambda = 5$.

Summary

$$PMF(x) = \frac{\exp(-\lambda)\,\lambda^x}{x!}$$

$$F(x) = \sum_{i=0}^{x} \frac{\exp(-\lambda)\lambda^i}{i!}$$

$$E(X) = \lambda$$

$$V(X) = \lambda$$

Example 6.5 Suppose the number of defects in typewriters coming off a production line is monitored over a period of time and the mean number of defects is found to be 0.5 per typewriter. Determine the probability of 0 and 1 defects being found.

SOLUTION

$$P(X = 0|0.5) = \frac{\exp(-0.5)0.5^0}{0!} = \frac{0.606 \times 1}{1} = 0.606$$

$$P(X = 1|0.5) = \frac{\exp(-0.5)0.5^1}{1!} = \frac{0.606 \times 0.5}{1} = 0.303$$

Example 6.6 In a large electrical network system the average number of cable faults per year per 100 km of cable is 0.5. Consider a specified piece of cable 10 km long and evaluate the probabilities of 0, 1, 2 etc. faults occurring in:

(a) A 20-year period;
(b) A 40-year period.

SOLUTION Assuming the average failure rate data to be valid for the 10 km cable and for the two periods being considered, the expected failure rate λ is

$$\lambda = \frac{0.5 \times 10}{100} = 0.05 \quad \text{faults/year}$$

(a) For a 20-year period,

$$E(x) = 0.05 \times 20 = 1.00,$$

and

$$P(x) = \frac{(\exp(-1)1^x)}{xi} \quad \text{for } x = 0, 1, 2 \ldots$$

(b) For a 40-year period,

$$E(x) = 0.05 \times 40 = 2.00,$$

and

$$P(x) = \frac{(\exp(-2)2^x)}{x!}, \quad \text{for } x = 0, 1, 2 \ldots$$

The Poisson distribution can be derived as a limiting form of the binomial if the following three assumptions are simultaneously satisfied:

1. n becomes large (that is, $n \to \infty$)
2. p becomes small (that is, $p \to 0$)
3. np remains constant.

Under these conditions, the binomial distribution with parameters n and p can be approximated to the Poisson distribution with parameter $\lambda = np$. This means that the Poisson distribution provides a good approximation to the binomial distribution if p is close to zero and n is large. Since p and q can be interchanged by simply interchanging the definitions of success and failure, the Poisson distribution is also a good approximation when p is close to one and n is large.

As an example of the use of the Poisson distribution as an approximation to the binomial distribution, the case in which $n = 10$ and $p = 0.10$ will be considered. The Poisson parameter for the approximation is then $\lambda = np = 10 \times 0.10 = 1$. The binomial distribution and the Poisson approximation are shown in Table 6.2. The two distributions agree reasonably well. If more precision is desired, a possible rule of thumb is that the Poisson is a good approximation to the binomial if $n/p > 500$ (this should give accuracy to at least two decimal places).

6.2 CONTINUOUS PROBABILITY DISTRIBUTIONS

This type of probability distribution is related to the continuous random variable. In the theory of probability there are several different rules for distribution functions which are derived in a purely theoretical way. The most frequently used rules for distribution functions in engineering practice

Table 6.2 Poisson distribution as an approximation to the binomial distribution

	Binomial $P(X = x\|n = 10, p = 0.1)$	Poisson $P(X = x\|\lambda = 1)$
0	0.598737	0.606531
1	0.315125	0.303265
2	0.074635	0.075816
3	0.010475	0.012636
4	0.000965	0.001580
5	0.000061	0.000158
6	0.000003	0.000013
7	0.000000	0.000001
8	0.000000	0.000000
9	0.000000	0.000000
10	0.000000	0.000000

are exponential, normal, lognormal, and Weibull. Each of them will be discussed in this chapter.

Each of these rules defines a family of distribution functions. Each member of the family is defined with a few parameters which in their own way control the distribution. All parameters can be classified in the following three categories:

1. Scale parameter, A, which defines the location of the distribution on the horizontal scale.
2. Shape parameter, B, which controls the shape of the distribution curves.
3. Source parameter, C, which defines the origin or the minimum value which a random variable can have.

Thus, individual members of a specific family of the probability distribution are defined by fixing numerical values for these parameters.

6.2.1 Exponential distribution

This type of probability distribution is fully defined by a single parameter which governs the location of the distribution, as well as its shape. Thus, according to the notation introduced above, $A = B$. The mathematical expression for the probability density function, in the case of the exponential distribution, is defined by the following rule

$$f(x) = \frac{1}{A} \exp\left(-\left(\frac{x}{A}\right)\right), \quad x > 0 \tag{6.9}$$

In Fig. 6.4 several graphs are shown of exponential density functions with

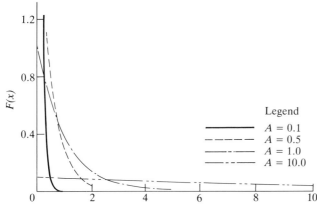

Figure 6.4 *PDF* for the exponential distribution.

different values of A. Notice that the exponential distribution is positively skewed, with the mode occurring at the smallest possible value, zero.

Making use of the general expression for the cumulative distribution function in the case of the continuous random variables, the above expression will have the following form after the integration

$$F(x) = P(X < x) = 1 - \exp\left(-\left(\frac{x}{A}\right)\right) \tag{6.10}$$

In Fig. 6.5 several curves of cumulative distribution functions are shown for different values of A. It can be shown that the mean and variance of the exponential distribution are

$$E(X) = M = A \tag{6.11}$$
$$V(X) = A^2 \tag{6.12}$$

The standard deviation in the case of the exponential distribution rule has a numerical value identical to the mean and the scale parameter, $SD(X) = E(X) = A$.

Summary

$$f(x) = \frac{1}{A}\exp\left(-\left(\frac{x}{A}\right)\right)$$
$$F(x) = 1 - \exp\left(-\left(\frac{x}{A}\right)\right)$$
$$E(X) = A$$
$$V(X) = A^2$$

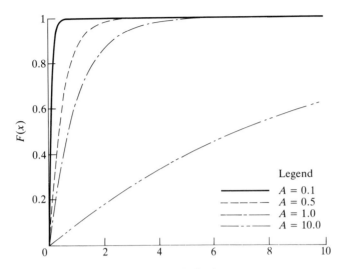

Figure 6.5 *CDF* of the exponential distribution.

Example 6.7 On average a machine breaks down once in 10 days. Find the chance that less than five days will elapse between breakdowns.

SOLUTION Based on the available data we are dealing with a probability distribution which is governed by the exponential rule where the scale parameter A is 10 days. Accordingly, the probability density function is $f(x) = 1/10 \exp(-x/10)$ and the required chance is $P(X < 5)$. The required probability is determined by the value of the *CDF* where $x = 5$, thus

$$P(X < 5) = F(5) = \int_0^5 \frac{1}{10} \exp\left(-\left(\frac{x}{10}\right)\right) dx = 1 - \exp\left(\frac{-5}{10}\right) = 1 - 0.61 = 0.39$$

Example 6.8 Suppose that the length of a phone call in minutes is an exponential random variable with parameter $A=3$. If someone arrives immediately ahead of you at a public telephone box, find the probability that you will have to wait (a) more than three minutes, and (b) between three and six minutes.

SOLUTION Letting X denote the length of the call made by the person in the phone box the desired probabilities are

(a) $$P(X > 3) = \int_3^\infty 1/3 \exp\left(-\left(\frac{x}{3}\right)\right) dx = \exp(-1) = 0.368$$

(b) $$P(3 < X < 6) = \int_3^6 1/3 \exp\left(-\left(\frac{x}{3}\right)\right) dx = \exp(-1) - \exp(-2) = 0.233$$

6.2.2 Normal distribution

This is the most frequently used and most extensively covered theoretical distribution in the literature. The normal distribution is continuous for all values of X between $-\infty$ and $+\infty$. It has a characteristic symmetrical shape, which means that the mean, the median and the mode have the same numerical value. The mathematical rule for its probability density function is as follows

$$f(x) = \frac{1}{\sqrt{2\pi}B} \exp \left[-\frac{1}{2} \left(\frac{x-A}{B} \right)^2 \right] \tag{6.13}$$

where A is a scale parameter and B is a shape parameter.

This rule pairs a probability density $f(x)$ with each and every possible value of X. The influence of the parameter A on the location of the distribution on the horizontal axis is shown in Fig. 6.6, where the values for parameter B are constant. Several different shapes of the probability distribution are obtained by varying the value of B while parameter A is kept constant, as shown in Fig. 6.7.

As a deviation of x from the scale parameter A enters as a squared quantity, *two* different x values, showing the same absolute deviation from A, will have the same probability density according to this rule. This dictates the symmetry of the normal distribution. Parameter A can be any finite number, while B can be any *positive* finite number.

The cumulative distribution function for the normal distribution is

$$F(a) = P(X \le a) = \int_{-\infty}^{a} f(x)dx$$

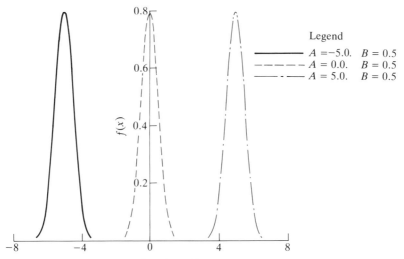

Figure 6.6 Probability density function with B constant.

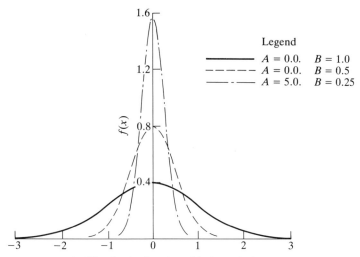

Figure 6.7 Probability density function with A constant.

where $f(x)$ is the normal density function. Taking into account Eq. (6.12) this becomes

$$F(a) = \int_{-\infty}^{a} \frac{1}{\sqrt{2\pi}B} \exp\left[-\frac{1}{2}\left(\frac{a - A}{B} \right)^2 \right] dx \qquad (6.14)$$

In Figs 6.8 and 6.9 several cumulative distribution functions are given of the normal distribution, corresponding to different values of A and B.

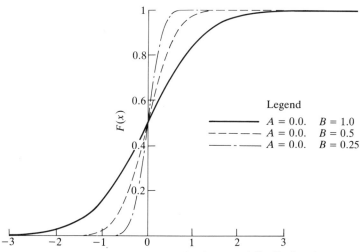

Figure 6.8 Cumulative distribution function for normal distribution A.

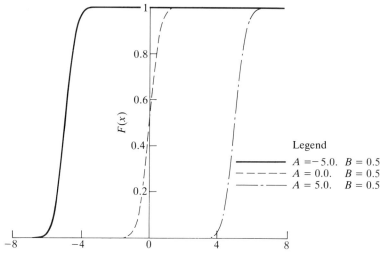

Figure 6.9 Cumulative distribution function for normal distribution B.

As the integral in Eq. (6.14) cannot be evaluated in a closed form, statisticians have constructed the table of probabilities which complies with the normal rule for the standardized random variable, Z. This is a theoretical random variable with parameters $A = 0$ and $B = 1$. The relationship between standardized random variable Z and random variable X is established by the following expression

$$z = \frac{x - A}{B} \tag{6.15}$$

Making use of this expression, Eq. (6.13) becomes simpler

$$f(z) = \frac{1}{\sqrt{2\pi}} \exp\left(-\frac{1}{2}z^2\right) \tag{6.16}$$

The standardized form of the distribution makes it possible to use one table only for the determination of *PDF* for any normal distribution, regardless of its particular parameters (see Table T1 on page 222).

The relationship between $f(x)$ and $f(z)$ is

$$f(x) = \frac{f(z)}{B} \tag{6.17}$$

By substituting $\dfrac{x - A}{B}$ with z Eq. (6.14) becomes

$$F(x) = \int_{-\infty}^{z} \frac{1}{\sqrt{2\pi}} \exp\left(-\frac{1}{2}z^2\right) dz = \Phi\left(\frac{x - A}{B}\right) \tag{6.18}$$

where Φ is the standard normal distribution function defined by

$$\Phi(z) = \int_{-\infty}^{x} \frac{1}{\sqrt{2\pi}} \exp\left(-\frac{1}{2}z^2\right) dx \qquad (6.19)$$

The corresponding standard normal probability density function is

$$f(z) = \frac{1}{\sqrt{2\pi}} \exp\left(\frac{-z^2}{2}\right) \qquad (6.20)$$

Most tables of the normal distribution give the cumulative probabilities for various *standardized* values. That is, for a given z value, the table provides the cumulative probability up to, and including, that standardized value in a normal distribution.

The expectation of a random variable, in the case of the normal distribution, is equal to the scale parameter A thus

$$E(X) = A \qquad (6.21)$$

whereas the variance is

$$V(X) = B^2 \qquad (6.22)$$

Summary

$$f(x) = \frac{1}{\sqrt{2\pi}B} \exp\left[-\frac{1}{2}\left(\frac{x-A}{B}\right)^2\right]$$

$$F(x) = \int_{-\infty}^{x} \frac{1}{\sqrt{2\pi}B} \exp\left[-\frac{1}{2}\left(\frac{x-A}{B}\right)^2\right] dx$$

$$E(X) = A$$

$$V(X) = B^2$$

Example 6.9 The catering department installed 2000 coffee machines which have an average life of 2000 hours with a standard deviation of 400 hours.

(a) How many machines might be expected to fail in the first 1000 operating hours?
(b) What is the probability of a coffee machine failing between 1800 and 2400 operating hours?
(c) After how many hours would one expect 5 per cent of the machines to have failed?

SOLUTION The operating life of this particular machine can be represented with a random variable, say T, whose probability distribution is defined as $N(2000, 400)$.

(a) $P(T < 1000) = F(1000) = ?$

According to Eq. (6.15) $z = (1000 - 2000)/400 = -2.5$. From Table T1 (given in the appendix) the required probability is $\Phi(-2.5) = 0.00621$. Thus, the expected number of failed coffee machines is: $2000 \times 0.00621 = 12.4 = 12$.

(b) $P(1800 < T < 2400) = F(2400) - F(1800) = ?$

The required probability can be determined by making use of Eq. (6.18), thus: $F(2400) = \Phi(1.0) = 0.84135$ and $F(1800) = \Phi(-0.5) = 0.30848$. Therefore, the probability of a coffee machine failing in the specified interval of operating time is $P(1800 < T < 2400) = 0.53287$.

(c) Here we have an inverse problem: to determine the value of t that corresponds to a probability of 0.05, $P(T < t) = 0.05$. According to Table T1 the cumulative distribution function, $F(z)$ for $z = -1.64$ corresponds to the specified probability. Thus, the task is to find the numerical value of t for which $z = -1.64$. Making use of Eq. (6.15) the solution becomes

$$\frac{(t - 2000)}{400} = -1.64 \rightarrow t = 1344 \text{ hours.}$$

6.2.3 Lognormal distribution

The lognormal probability distribution can in some respects be considered as a special case of the normal distribution because of the derivation of its probability function. If a random variable $Y = \ln X$ is normally distributed $N(A_Y, B_Y)$ then the random variable X follows the lognormal distribution. Thus, the probability density function for a random variable X is defined as

$$f_X(x) = \frac{1}{x B_Y \sqrt{2\pi}} \exp\left[-\frac{1}{2} \left(\frac{\ln x - A_Y}{B_Y} \right)^2 \right] \quad x \geqslant 0 \qquad (6.23)$$

As in the case of normal distribution the parameter A_Y is called the *scale parameter* (see Fig. 6.10) and parameter B_Y is called the *shape parameter* (see Fig. 6.11).

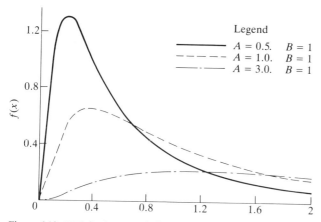

Figure 6.10 *PDF* for lognormal distribution with constant shape parameter.

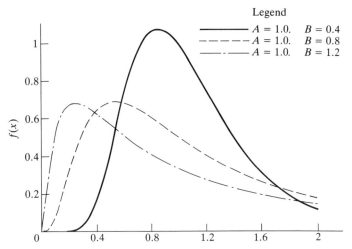

Figure 6.11 *PDF for lognormal distribution with constant scale parameter.*

The relationship between parameters A_X and A_Y is defined

$$A_X = \exp(A_Y + \frac{1}{2}B_Y^2) \tag{6.24}$$

and the relationship between B_X and B_Y is

$$B_X = \sqrt{A_X^2(\exp(B_Y^2) - 1)} \tag{6.25}$$

The cumulative distribution function for the lognormal distribution is defined as

$$F_X(x) = (X \leq x) = \int_0^x \frac{1}{xB_Y\sqrt{2\pi}} \exp\left[-\frac{1}{2}\left(\frac{\ln x - A_Y}{B_Y}\right)^2\right] dx \tag{6.26}$$

As the integral cannot be evaluated in closed form the same procedure is applied as in the case of normal distribution. Then, making use of the standardized random variable Eq. (6.26) transforms into

$$F_X(x) = P(X \leq x) = \Phi\left(\frac{\ln x - A_Y}{B_Y}\right) \tag{6.27}$$

where Φ is the standard normal distribution function whose numerical values can be found in Table T1, for

$$z = \frac{\ln x - A_Y}{B_Y} \tag{6.28}$$

Several cumulative distribution functions are shown in Figs 6.12 and 6.13.

The measures of central tendency in the case of lognormal distributions are defined by the location parameters and the variance

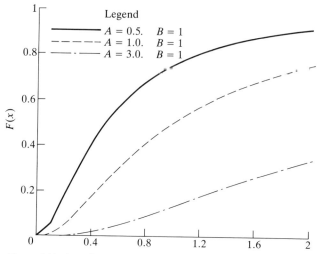

Figure 6.12 *CDF* for lognormal distribution with constant shape parameter.

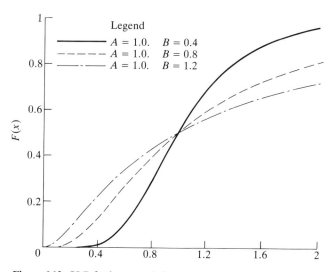

Figure 6.13 *CDF* for lognormal distribution with constant scale parameter.

- Mean

$$M = E(X) = \exp(A_Y + \frac{1}{2}B_Y^2)$$ (6.29)

- Median

$$Md = \exp(A_Y)$$ (6.30)

- Mode

$$Mo = \exp(A_Y - B_Y^2)$$ (6.31)

- Deviation parameter (the variance)

$$V(X) = \exp(2A_Y + B_Y)^2 \left[\exp(B_Y^2 - 1)\right]$$ (6.32)

Summary

$$f_X(x) = \frac{1}{xB_Y\sqrt{2\pi}} \exp\left[-\frac{1}{2}\left(\frac{\ln x - A_Y}{B_Y}\right)^2\right]$$

$$F_X(x) = \Phi\left(\frac{\ln x - A_Y}{B_Y}\right)$$

$$E(X) = \exp\left(A_Y + \frac{1}{2}B_Y^2\right)$$

$$V(X) = \exp\left(2A_Y + B_Y\right)^2 \left[\exp\left(B_Y^2 - 1\right)\right]$$

So far the range of random variables considered has been between 0 and + infinity. The lognormal distribution can be successfully applied in the cases where the random variable has a range, $C < X < +$ infinity, where C can take any value greater than zero. As has already been pointed out the parameter C is called a source parameter and represents a value below which probability is equal to 0. This type of lognormal distribution is known as the *three-parameter distribution*. The relevant equations for the three-parameter lognormal distribution are given below

$$f_X(x) = \frac{1}{xB_Y\sqrt{2\pi}} \exp\left[-\frac{1}{2}\left(\frac{\ln(x-C) - A_Y}{B_Y}\right)^2\right]$$

$$F_X(x) = \Phi\left(\frac{\ln(x-C) - A_Y}{B_Y}\right)$$

$$E(X) = C + \exp\left(A_Y + \frac{1}{2}B_Y^2\right)$$

$$V(X) = C + \exp\left(2A_Y + B_Y\right)^2 \left[\exp\left(B_Y^2 - 1\right)\right]$$

Example 6.10 The age of motor vehicles can be represented by a lognormal distribution with parameters $A = 1.75$ and $B = 0.57$. What percentage of the vehicles are less than five years old?

SOLUTION

$$P(\text{Age of motor vehicle} < 5) = F(5) = \Phi\left(\frac{\ln(5) - 1.75}{0.57}\right) = 0.40$$

Thus, 40 per cent of vehicles are less than five years old.

Example 6.11 If we are interested in the age distribution of motor vehicles in the UK which possess a valid MOT certificate, the distribution function will be defined by three parameters because the certificate is not needed for vehicles younger than three years. Thus, the random variable T is defined as $LN(1.75, 0.57, 3)$. What percentage of the vehicles are less than five years old?

SOLUTION

$$P(\text{Age of motor vehicle with certificate} < 5) = F(5) = \Phi\left(\frac{\ln(5-3) - 1.75}{0.57}\right) = 0.03$$

Thus less than 30 per cent of vehicles with a valid MOT certificate are less than five years old.

6.2.4 Weibull distribution

This distribution originated from the experimentally observed variations in the yield strength of Bofors steel, the size distribution of fly ash, fibre strength of Indian cotton, and the fatigue life of a $St-37$ steel by the Swedish engineer W. Weibull. As the Weibull distribution has no characteristic shape, such as the normal distribution, for example, it has a very important role in the statistical analysis of experimental data. The shape of this distribution is governed by its parameters.

The rule for the probability density function of the Weibull distribution is

$$f(x) = \frac{B}{A-C}\left(\frac{x-C}{A-C}\right)^{B-1} \exp\left[-\left(\frac{x-C}{A-C}\right)^{B}\right] \tag{6.33}$$

where $A > 0$, $B > 0$ and $C > 0$. As the parameter C is often set equal to zero this can be rewritten

$$f(x) = \frac{B}{A}\left(\frac{x}{A}\right)^{B-1} \exp\left[-\left(\frac{x}{A}\right)^{B}\right] \tag{6.34}$$

By altering parameter B the Weibull distribution takes different shapes, as illustrated in Figs 6.14 and 6.15. For example, when $B = 3.4$ the Weibull approximates to the normal distribution; when $B = 1$, it is identical to the exponential distribution.

The cumulative distribution function for the Weibull distribution is

$$F(x) = 1 - \exp\left[-\left(\frac{x-C}{A-C}\right)^{B}\right] \tag{6.35}$$

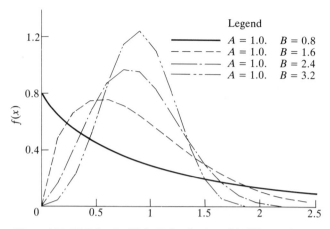

Figure 6.14 *PDF* for the Weibull distribution with different shape parameters.

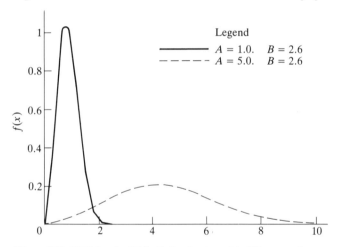

Figure 6.15 *PDF* for the Weibull distribution with different scale parameters.

or

$$F(x) = 1 - \exp\left[-\left(\frac{x}{A}\right)^B\right] \qquad (6.36)$$

Figure 6.16 shows the *CDF* for the Weibull distribution with different shape parameters and Fig. 6.17 shows the *CDF* with different scale parameters.

The expected value of the Weibull distribution is given by

$$E(X) = (A - C) \times \Gamma\left(\frac{1}{B} + 1\right) \qquad (6.37)$$

or

$$E(X) = A \times \Gamma\left(\frac{1}{B} + 1\right) \qquad (6.38)$$

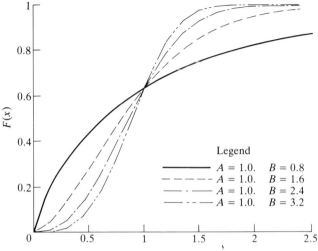

Figure 6.16 *CDF* for the Weibull distribution with different shape parameters.

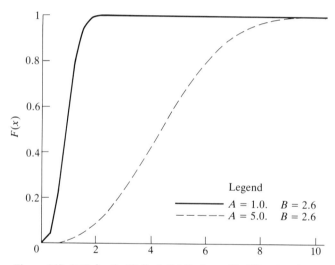

Figure 6.17 *CDF* for the Weibull distribution with different scale parameters.

where Γ is the gamma function.

The variance of the Weibull distribution is given by

$$V(X) = (A - C)^2 \left[\Gamma \left(1 + \frac{2}{B} \right) - \Gamma^2 \left(1 + \frac{1}{B} \right) \right] \qquad (6.39)$$

or

$$V(X) = A^2 \left[\Gamma \left(1 + \frac{2}{B} \right) - \Gamma^2 \left(1 + \frac{1}{B} \right) \right] \qquad (6.40)$$

Numerical values for the gamma function are given in Table T4 (page 282) for $0.5 < B < 20$.

Summary

$$f(x) = \frac{B}{A-C}\left(\frac{x-C}{A-C}\right)^{B-1}\exp\left[-\left(\frac{x-C}{A-C}\right)^{B}\right]$$

$$\frac{B}{A}\left(\frac{x}{A}\right)^{B-1}\exp\left[-\left(\frac{x}{A}\right)^{B}\right]$$

$$F(x) = 1 - \exp\left[-\left(\frac{x-C}{A-C}\right)^{B}\right]$$

$$1 - \exp\left[-\left(\frac{x}{A}\right)^{B}\right]$$

$$E(X) = (A-C)\times\Gamma\left(\frac{1}{B}+1\right)$$

$$A\times\Gamma\left(\frac{1}{B}+1\right)$$

$$V(X) = (A-C)^2\left[\Gamma\left(1+\frac{2}{B}\right)-\Gamma^2\left(1+\frac{1}{B}\right)\right]$$

$$A^2\left[\Gamma\left(1+\frac{2}{B}\right)-\Gamma^2\left(1+\frac{1}{B}\right)\right]$$

Example 6.12 Assuming that the operational life of a certain component can be represented by the Weibull distribution with $B=4$, $A=2000$, and $C=1000$, find the probability that the component will not fail in the first 1500 hours.

SOLUTION The required probability can be calculated by applying Eq. (6.35), thus

$$P(X > 1500) = 1 - P(X \le 1500) = 1 - F(1500) = \exp\left[-\left(\frac{1500-1000}{2000-1000}\right)^4\right] = 0.939$$

SEVEN

MEASURES OF RELIABILITY

The main objective of this chapter is to define the measures by which reliability can be numerically defined and described, just as the functionability of the item has been numerically described through the different parameters, under the general heading of performance.

Since it is readily accepted that a population of supposedly identical copies of an item will fail at different instants of time, it follows that timing of failure phenomena can only be described in probabilistic terms. In reliability theory the following three approaches are recognized:

1. Time to failure
2. Stress-strength
3. Condition based.

All will be presented and examined in this chapter.

7.1 TIME TO FAILURE APPROACH TO RELIABILITY

In order to numerically express the reliability of an item, according to this approach, it is necessary to establish a relationship between the concept of reliability (presented in Chapter 2) and the concept of the probability system (given in Chapter 5). In these chapters it was shown how the process of maintaining the functionability of an item during its operational life can be considered as a random experiment, and how the transition to a state of failure can be regarded as an elementary event which represents its outcome.

The function which assigns a corresponding numerical value t_i to every elementary event (failure) a_i from sample space S, is a random variable which will be called the time to failure, *TTF*, as shown in Fig. 7.1. Thus the probability that random variable *TTF* will take on a value t_i is denoted by $p_i = P(T = t_i)$. The numerical values taken on by random variables and the probability of their occurrence define the probability distribution.

This establishes the complete analogy between the probability system and the ability of an item to maintain its functionability. It should be noted at this point that although time is the true variable, it may often be more convenient to use other readily available variables which represent time of use such as miles read from an odometer, hours of operation, number of cycles, number of switches, total accumulated energy, etc. Thus the concept of reliability has a close relationship with probability. By using probability theory the hitherto qualitative characteristics concerning the reliability of the item such as very reliable, quite reliable, and similar terms, can be translated into quantitative measures.

According to this approach reliability can be fully defined by the probability distribution of the random variable *TTF*. As already pointed out in Chapter 5, and as shown by Fig. 7.1, the probability distribution of any random variable can be defined through the probability functions (density function, cumulative function), or the summary measures (expected value, median, variance and similar).

The measures of the probability system which are applied to the field of reliability are:

1. Failure function
2. TTF_p life
3. Reliability function
4. Expected time to failure
5. Mission success
6. Hazard function.

A brief definition and description of these characteristics follows, together with numerical examples to illustrate the main points.

7.1.1 Failure function

As the cumulative probability function represents the probability that the random variable under consideration will be equal to or less than a particular value, in the case of the random variable *TTF*, this will be called the *failure function*, $F(.)$. Its numerical value presents the probability that the transition to a state of failure (or simply its failure), will occur before or at the moment of operating time, t, thus

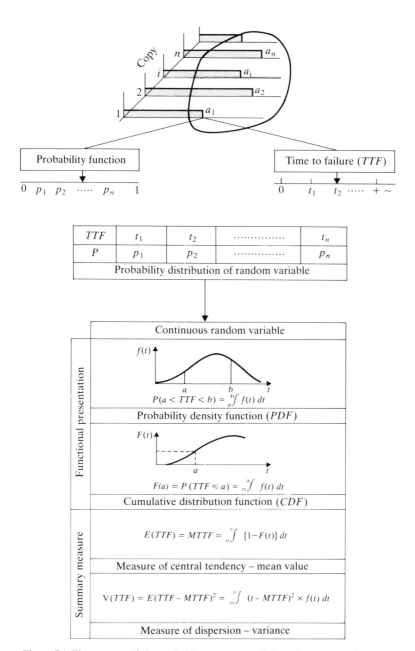

Figure 7.1 The concept of the probability system applied to the concept of reliability.

Table 7.1 Failure function, $F(t)$ for well-known theoretical distributions

Exponential	$1 - \exp\left(-\left(\frac{t}{A}\right)\right)$ $t > 0$
Normal	$\Phi[(t - A)/B]$ $-\infty < t < +\infty$
Lognormal	$\Phi[(\ln(t) - A)/B]$ $t > 0$
Weibull	$1 - \exp\left[-\left(\frac{(t-C)}{(A-C)}\right)^{B}\right]$ $t > 0$

$$F(t) = P(\text{failure will occur before or at time } t)$$
$$= P(TTF \leq t)$$
$$= \int_0^t f(t)dt \tag{7.1}$$

where $f(t)$ is the probability density function of TTF. Table 7.1 shows the failure function for some well-known theoretical distributions.

It should be noted that in the case of the normal distribution the failure function exists from $-\infty$ and so may have significant value at $t = 0$. Since negative time is usually meaningless in reliability, great care should be used in manipulating this model unless $A > 3 \times B$, when the intercept at $t = 0$ can be considered negligible.

Example 7.1 An item on a commercial aircraft has an expected operating life of 1000 hours. Assume an exponential distribution of the operating time to failure and answer the following questions:

(a) What is the probability of failure during a four-hour flight?
(b) What is the maximum length of flight such that the probability of survival will not be less than 0.99 (assume that the item is in continuous operation during flight)?

SOLUTION

(a) $$P(TTF \leq 4) = F(4) = \int_0^4 f(t)dt$$

for exponential distribution it will be

$$F(4) = 1 - \exp[-(4/1000)] = 0.00399$$

(b) $$P(TTF > t) = 0.99 = \exp[-(t/1000)], \quad \longrightarrow t = 10.05 \text{ hours}$$

Maximum length of flight providing the required probability of survival of 0.99 is 10 hours.

Example 7.2 Operation time to failure of an item with a normal distribution can be presented by $A = 20\,000$ cycles and $B = 2000$. Determine the probability that the item will fail before the completion of $21\,000$ cycles.

SOLUTION The solution to this problem can be found by applying the expression for the failure function for the normal distribution given in Table 7.1. Thus

$$P(TTF < 21\,000) = F(21\,000) = \Phi[(21\,000 - 20\,000)/2000] = 0.69148$$

Example 7.3 A large number of identical relays have times to first failure that follow a Weibull distribution with shape parameter $B = 0.5$ and scale parameter $A = 16$ years. What is the probability that a relay will fail during (a) one year, (b) five years and (c) ten years of operation?

SOLUTION
(a) $F(1) = P(TTF < 1) = 1 - \exp(-(1/16)^{0.5}) = 0.2212$
(b) $F(5) = P(TTF < 5) = 1 - \exp(-(5/16)^{0.5}) = 0.4282$
(c) $F(10) = P(TTF < 10) = 1 - \exp(-(10/16)^{0.5}) = 0.5464$

7.1.2 · TTF_p life

This measure of reliability is related to the length of operating time by which a given percentage p of a population will have failed. Numerically, it is the abscissa of the intersection point between the cumulative distribution function and the horizontal line originating from the point or ordinate which corresponds to the chosen percentage of failure. The most frequently used are TTF_{10}, TTF_{50} and TTF_{90} which represent the length of operational life up to which 10, 50 and 90 per cent of population failed, see Fig. 7.2.

Mathematically, TTF_p life can be represented in the following way

$$TTF_p = t \rightarrow \text{for which } F(t) = P(TTF \le t) = \int_0^t f(t)dt = p \qquad (7.2)$$

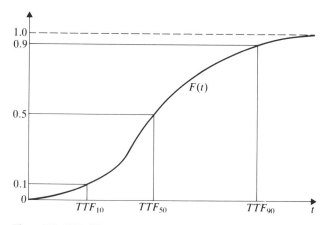

Figure 7.2 TTF_p life.

If we are interested in the operating time up to which, say, 10 per cent of the population will lose functionability, TTF_{10}, Eq. (7.2) will take the following form

$$TTF_{10} = t \rightarrow \text{for which } F(t) = P(TTF \leq t) = \int_0^t f(t)dt = 0.1$$

Example 7.4 A company has installed 1745 coffee machines which have an expected life of 1145 hours with a standard deviation of 200 hours.

(a) How many machines may be expected to fail between 900 and 1300 burning hours?
(b) After what period of operating time would we expect 10 per cent of them to have failed, or what is TTF_{10} life?

SOLUTION

(a) $\quad P(900 \leq TTF \leq 1300) = P(TTF \leq 1300) - P(TTF \leq 900)$

$$= F(1300) - F(900)$$

$$= \Phi\left(\frac{1300 - 1145}{200}\right) - \Phi\left(\frac{900 - 1145}{200}\right)$$

$$= 0.7798 - 0.1112 = 0.6676$$

The company can expect 66.76 per cent of coffee machines to fail in this interval of operating time.

(b)

$$TTF_{10} = P(TTF < t) = 0.1, \rightarrow \Phi\left(\frac{t - 1145}{200}\right) = -1.28 \quad \longrightarrow \quad TTF_{10} = t = 889 \text{ hours.}$$

7.1.3 Reliability function

A function, $F'(.)$, which is complementary to the failure function, represents the probability that a random variable will have a value greater than some particular value. Thus, $F'(t) = 1 - F(t) = P(X > t)$. Applied to the concept of reliability, this function presents the probability that a state of functioning will be maintained up to a particular operating time t, and we will call it the *reliability function*, $R(t)$, thus:

$$R(t) = P(\text{state of functioning will be maintained up to time } t)$$

$$= P(TTF > t)$$

$$= \int_t^\infty f(t)dt \tag{7.3}$$

Since both functions, $F(t)$, $R(t)$, are defined by the integral of the probability density function $f(t)$, the probabilities of their occurrences are equal to the area under the given curve within given interval limits, and their numerical values are determined by the numerical value of t as shown in Fig. 7.3. Thus, for any value of t the equality $F(t) + R(t) = 1$ is valid. In the

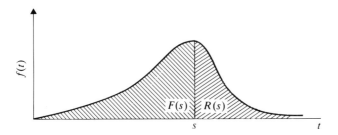

Figure 7.3 The relationship between the failure function and the reliability function.

literature the reliability function is sometimes called the *survival function,* for obvious reasons. Table 7.2 shows the reliability function for well-known theoretical distributions.

Example 7.5 If *TTF* is a random variable representing the hours to failure for the item under consideration and the corresponding probability density function is defined by the following expression

$$f(t) = t \times \exp\left(-\frac{t^2}{2}\right), \qquad t > 0$$

derive the expressions for the reliability function.

SOLUTION

$$R(t) = P(TTF > t) = \int_t^\infty f(t)dt$$

Table 7.2 Reliability function, $R(t)$ for well-known theoretical distributions

Exponential	$\exp\left(-\left(\frac{t}{A}\right)\right)$ $t > 0$
Normal	$\Phi[(A - t)/B]$ $-\infty < t < +\infty$
Lognormal	$\Phi[(A - \ln(t))/B]$ $t > 0$
Weibull	$\exp\left[-\left(\frac{(t-C)}{(A-C)}\right)^B\right]$ $t > 0$

$$= \int_t^\infty t \times \exp(-\frac{t^2}{2})dt$$

Substituting $z = t^2$, $dz = 2tdt$ or $dt = dz/2t$
we obtain

$$R(t) = \int_t^\infty \sqrt{z}\exp(-z/2)\frac{dz}{2\sqrt{z}}$$

$$= 1/2(-2\exp\left(-\frac{z}{2}\right)$$

$$= \exp\left(-\frac{t^2}{2}\right)$$

7.1.4 Expected time to failure

The expected time to failure of an engineering item can be determined by using the expectation of the random variable TTF, thus

$$E(TTF) = \int_0^\infty tf(t)dt \tag{7.4}$$

In the literature the expected time to failure is known as the *mean time to failure*, thus $E(TTF) = MTTF$, see Fig. 7.4 and Table 7.3.

Making use of Eq. (5.21) the same result can be obtained by taking the integral of the reliability function, because $R(t) = 1 - F(t)$, thus

$$E(TTF) = MTTF = \int_0^\infty R(t)dt \tag{7.5}$$

Example 7.6 A large number of identical relays have times to first failure that follow a Weibull distribution with shape parameter $B = 0.5$ and scale parameter $A = 16$ years. Determine the numerical value for the mean time to failure of the relay under consideration.

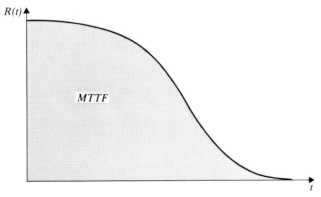

Figure 7.4 Diagrammatical representation of expected time to failure.

Table 7.3 Expected operating life, $E(TTF) = MTTF$ for well-known distributions

Exponential	A
Normal	A
Lognormal	$\exp(A)$
Weibull	$A \times \Gamma(1 + 1/B)$

SOLUTION According to the expression for expected value of *TTF* given in Table 7.3 the corresponding numerical value for the relay analysed will be:

$$MTTF = A\Gamma(1/B + 1) = 16 \times 2 = 32 \text{ hours}$$

where $\Gamma(1/0.5 + 1) = 2$, based on Table T4.

7.1.5 Mission success

The reliability function, defined by Eq. (7.3), represents the probability that the item under consideration will maintain functionability from the beginning of the operation up to time t. In many cases it is important to know the probability that an item, which is functionable at the present time, will maintain its functionability during a future *mission*. This is a typical conditional probability problem, because the item can maintain functionability throughout the mission given that it was functionable at the beginning. This type of reliability measure we shall call *mission success*, $MS(t_b, \text{mission})$. It is fully defined by the following expression:

$$MS(t_b, \text{mission}) = P(TTF > t_b + \text{mission}|TTF > t_b)$$

where t_b is the instant of time the mission begins.

For ease of manipulation of the data the symbol t_e will be introduced which represents $t_e = t_b + \text{mission}$. Making use of Eq. (7.3), which defines the reliability function, $R(t)$, and applying the principles of conditional probability, the above expression for mission reliability can be rewritten in the following form

$$MS(t_b, t_e) = P(TTF > t_e|TTF > t_b) = \frac{R(t_e)}{R(t_b)} \quad (7.6)$$

Therefore, mission success is a conditional probability which is fully defined by the ratio of the numerical values of the reliability function at the end and at the beginning of the mission.

In cases where the beginning of the mission coincides with the beginning of the operation, the mission success is equal to the reliability function, thus

$$MS(0, t_e) = \frac{R(t_e)}{R(0)} = R(t_e)$$

because $R(0) = 1$. This type of reliability measure provides very useful information for operation and maintenance engineers.

Example 7.7 What is the probability that a lorry of a particular make and type which has done 34 000 miles from new will successfully complete the mission of 5000 miles? Assume that the time to failure can be described by the Weibull distribution with a scale parameter of 49 500 miles and a shape parameter of 2.6.

SOLUTION The notation used will be $t_b = 34\,000$ miles, $t_e = 34\,000 + 5000 = 39\,000$ miles. According to Eq. (7.6) the required probability will be

$$MS(34\,000, 39\,000) = \frac{R_e(39\,000)}{R_b(34\,000)} = 0.85$$

where

$$R_b(34\,000) = \exp\left(-\left(\frac{34000}{49500}\right)^{2.6}\right) = 0.686$$

$$R_e(39\,000) = \exp\left(-\left(\frac{39000}{49500}\right)^{2.6}\right) = 0.584$$

Thus, there is probability of 0.85 that the lorry planned for the mission will be functionable at the end of the 5000-mile journey.

7.1.6 Hazard function

This is one of the most frequently used characteristics in engineering and is used to represent the probability that an item, which maintained function-ability up to instant t, will fail during a short interval of time in the future. In the terms of probability theory, failure rate is the conditional probability density of failure at the instant t with the hypothesis that the item has been functioning up to that instant. The probability of failure of an item in a given time interval (t_1, t_2) can be expressed

$$P(t_1 \leq TTF \leq t_2) = \int_{t_1}^{t_2} f(t)dt = \int_0^{t_2} f(t)dt - \int_0^{t_1} f(t)dt = F(t_2) - F(t_1)$$

The rate at which failures occur within a certain time interval (t_1, t_2) is called the *failure rate*. It is defined as the probability that a failure occurs in the specified interval (t_1, t_2) per unit time, given that it has not occurred prior to t_1, the beginning of the interval. Thus the failure rate is

$$P(t_1 < TTF < t_2 \mid TTF > t_1) = \frac{F(t_2) - F(t_1)}{(t_2 - t_1)R(t_1)} = \frac{R(t_1) - R(t_2)}{(t_2 - t_1)R(t_1)}$$

If the interval (t_1, t_2) is redefined as $(t, t + \Delta t)$, this becomes

$$P(t < TTF < t + \Delta t \mid TTF > t) = \frac{R(t) - R(t + \Delta t)}{\Delta t R(t)} = \frac{F(t + \Delta t) - F(t)}{\Delta t R(t)}$$

$$(7.7)$$

The rate in the above definition is expressed as failures per unit time; in reality the units might be kilometres, hours, etc., as previously discussed.

The *hazard function* is defined as a limit of the failure rate as the interval approaches zero. Thus the hazard function is the instantaneous failure rate. The hazard function $z(t)$ is defined by the following expression

$$z(t) = \lim_{\Delta t \to 0} \frac{R(t) - R(t + \Delta t)}{\Delta t R(t)} = \frac{1}{R(t)} \left(\frac{-d}{dt} R(t) \right) = \frac{f(t)}{R(t)} \quad (7.8)$$

The quantity $z(t)dt$ represents the probability that an item of age t will fail in the small interval of time t to $t + \Delta t$. The above equation shows that at $t = 0$, $z(0) = f(0)$ since $R(0) = 1$. The importance of the hazard function is:

(a) It indicates the change in the failure rate over the operating life of the item.
(b) It shows that the hazard function is a conditional function of the probability density function, the conditional relationship being the reliability function. In physical terms this relationship means that the probability density function permits the probability of failure to be evaluated for any period of time in the future, whereas the hazard function permits the probability of failure to be evaluated in the next time period, given that it has survived up to time t.

7.1.7 Hazard function defined by theoretical distributions

Making use of the analytical rules which define the well-known theoretical probability distributions, given in Chapter 6, and using Eq. (7.8), expressions for the hazard function will now be derived.

Exponential distribution The hazard function is defined as

$$z(t) = \frac{1}{A} \quad (7.9)$$

and is illustrated by Fig. 7.5. The exponential distribution is the most widely used distribution in life testing. The reason is that it is an appropriate model where failure of an item is due not to deterioration as a result of wear, but rather to random events. However, it is necessary to emphasize that the failure process has to be random, and in accordance with the postulates of the Poisson process, which means that:

(a) The process is time homogeneous and the occurrence of random events is independent of past occurrences.
(b) The probability of exactly one random event in any time interval is proportional to the length of that time interval.

In other words, if the number of failures occurring over an interval of time

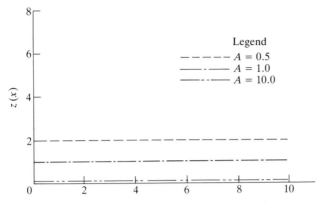

Figure 7.5 Hazard function for the exponential distribution.

is a Poisson distribution, then the time between failure is an exponential distribution.

Normal distribution The hazard function in case of normal distribution has the following expression

$$z(t) = \frac{f(t)}{R(t)} \tag{7.10}$$

Figure 7.6 shows the shape for different A values, and Fig. 7.7 for different B values. This type of probability distribution is extensively used in cases where the transition to $SoFa$ results from degradation of the materials from which the item is made. It covers deterioration from wear-out, corrosion, abrasion and similar processes.

Lognormal distribution The hazard function in the case of the lognormal distribution is defined by the expression

$$z(t) = \frac{f(t)}{R(t)} \tag{7.11}$$

Figures 7.8 and 7.9 on page 81 show several functions for different values of A and B. It is interesting to observe that the hazard function shows a tendency to decrease at the latter stage of the operation.

Weibull distribution The hazard function for the Weibull distribution is defined by the following equation

$$z(t) = \frac{B(t - C)^{B-1}}{(A - C)^B} \tag{7.12}$$

which is illustrated by Figs 7.10 and 7.11 on page 82. The ability of the Weibull distribution to model failure situations where non-constant hazard functions apply, makes this distribution one of the most generally useful distributions for analysing failure data.

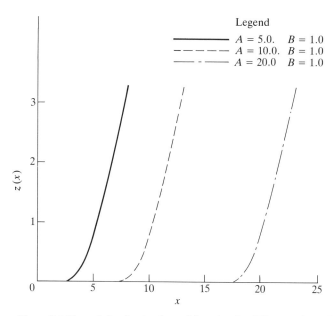

Figure 7.6 Normal distribution hazard function for different values of A.

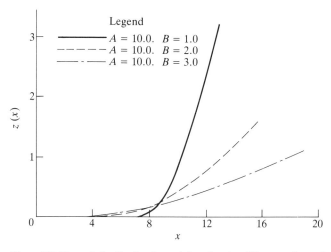

Figure 7.7 Normal distribution hazard function for different values of B.

If $B > 1$, this indicates an increasing $z(t)$ which may be symptomatic of wear out. For $B = 1, z(t)$ is constant, which means that occurrences of failure are random and in accordance with the postulates of the Poisson process; the resulting distribution is exponential. If $B < 1$, a decreasing $z(t)$ is indicated.

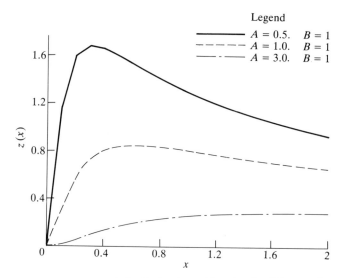

Figure 7.8 Lognormal distribution hazard function for different values of A.

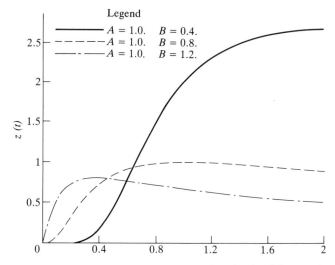

Figure 7.9 Lognormal distribution hazard function for different values of B.

Example 7.8 An XM-1 tank has a $MTTF$ of 810 miles. Assuming an exponential distribution determine the hazard function at 100, 200, 400, and 800 miles. What conclusion can be made regarding the behaviour of the hazard function?

SOLUTION According to Eq. (7.5) the hazard function at specified instances of time will be

$z(100) = 1/810 = 0.0012$
$z(200) = 1/810 = 0.0012$

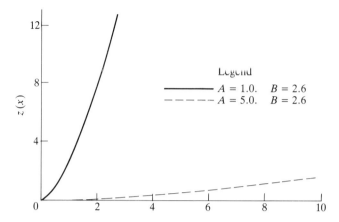

Figure 7.10 Weibull distribution hazard function for different values of A.

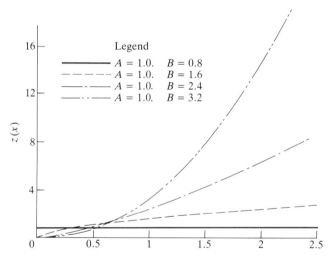

Figure 7.11 Weibull distribution hazard function for different values of B.

$$z(400) = 1/810 = 0.0012$$
$$z(800) = 1/810 = 0.0012$$

The hazard function has a constant value, i.e. it is independent of the mission length.

Example 7.9 A component has a normal distribution of operating time to failure with $A = 20\,000$ cycles and $B = 2000$ cycles. What is the failure rate after 18 000 cycles?

SOLUTION

$$P(TTF > 18000) = R(18\,000) = \int_{18\,000}^{\infty} f(t)dt$$

$$= 1 - \Phi\left(\frac{(18\,000 - 20\,000)}{2000}\right)$$

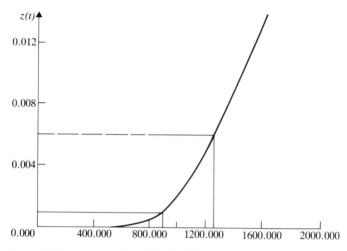

Figure 7.12 Hazard function, N (1145, 200).

$$= 1 - 0.1587 = 0.8413$$

$$f(18000) = \frac{f(z)}{2000}, \quad \text{where } z = \frac{18\,000 - 20\,000}{2000}$$

from Table T2 $f(z) = f(-1) = 0.24193$

$$z(18\,000) = \frac{f(18\,000)}{R(18\,000)} = \frac{f(z)/B}{R(18\,000)} = \frac{0.24193/2000}{0.8413} = 0.000\,1437$$

Example 7.10 An electric lamp has an expected life of 1145 burning hours with a standard deviation of 200 hours. Calculate the hazard rate at 900 and 1300 burning hours and determine whether the failure rate increases or not.

SOLUTION

$$z(900) = \frac{f(900)}{R(900)} = \frac{0.00094183}{0.889973} = 0.001058$$

$$z(1300) = f(1300)/R(1300) = \frac{0.0014771}{0.219143} = 0.0067398$$

Fig. 7.12 confirms that the failure rate increases with operation time.

Example 7.11 If TTF is a random variable representing the hours to failure for a device where

$$f(t) = t \exp\left(-\frac{t^2}{2}\right), \quad t > 0$$

find the expressions for hazard function.

SOLUTION

$$R(t) = P(TTF > t) = \int_t^\infty f(t)dt$$

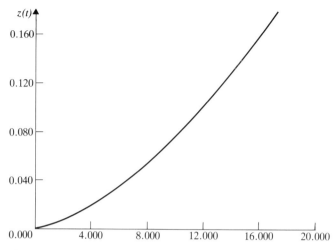

Figure 7.13 Hazard function for Example 7.12.

$$= \int_t^\infty t \exp\left(-\frac{t^2}{2}\right) dt$$

Substituting $z = t^2$, $dz = 2t\,dt$ or $dt = dz/2t$

$$R(t) = \int_{\sqrt{z}}^\infty \sqrt{z} \exp\left(-\frac{z}{2}\right) \frac{dz}{2\sqrt{z}}$$

$$= \frac{1}{2} \left[-2\exp(-z/2)\right] \Big|_{\sqrt{z}}^\infty$$

$$= \exp\left(-\frac{t^2}{2}\right)$$

$$z(t) = \frac{f(t)}{R(t)}$$

$$= t\frac{\exp(-t/2)}{\exp(-t/2)} = t$$

Example 7.12 A large number of identical relays have times to first failure that follow a Weibull distribution with shape parameter $B = 2.54$ and scale parameter $A = 16$ years. What is the failure rate at (a) one year, (b) five years and (c) ten years of operation?

SOLUTION
(a) $z(1) = 0.00222$, (b) $z(5) = 0.026471$ and (c) $z(10) = 0.076978$. Figure 7.13 shows the hazard function.

7.2 STRESS-STRENGTH APPROACH TO RELIABILITY

The basis of the stress-strength approach to reliability is to assume that a given item has a stress-resisting capability and that failure will occur if this is exceeded by the stresses induced under the operating conditions and regime. That is, item reliability can be expressed as a function of stress and strength distributions. This approach is based on two assumptions Barlow, (1965):

1. The strength of the item is independent of time (time-dependent features such as fatigue, creep and corrosion are neglected) Carter (1972).
2. If the stress due to the total load is less than the strength of the item then the state of functioning prevails, but if the stress is greater the item will fail.

Then, by definition:

$$SoFu = \text{strength greater than stress } (S > L)$$
$$SoFa = \text{strength less than stress } (S < L)$$

According to this approach, therefore, reliability is the probability that the strength is greater than stress, $R(Tst) = P(S > L, Tst)$. Taking into account the first assumption, reliability can be expressed as $R = P(S > L)$.

In order to compute reliability, the following notation will be used: S the random variable which represents the strength, $f_S(s)$ the probability density function for the stress, L the random variable which represents the stress, $f_L(s)$, as shown in Fig. 7.14.

Looking at this figure it is clear that reliability depends only on the *interference* of the stress and strength distributions. The shaded portion shows the interference area, which is *indicative* of the probability of failure, and is expanded on in Fig. 7.15.

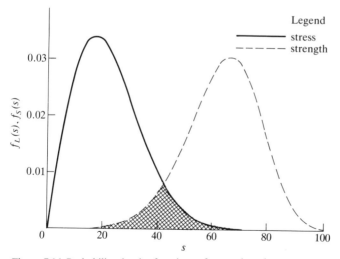

Figure 7.14 Probability density functions of strength and stress.

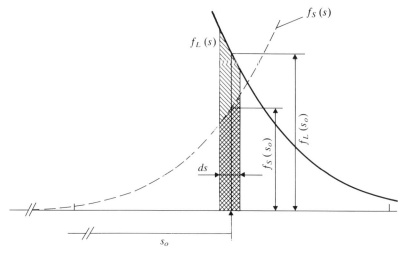

Figure 7.15 Interference area.

The probability that a random variable L will have a value lying in small interval ds is equal to the area of the element ds; that is

$$P(s_0 - ds/2 < L < s_0 + ds/2) = f_L(s_0)ds \qquad (7.13)$$

The probability that the strength S is greater than a certain stress s_0 is given by

$$P(S > s_0) = \int_{s_0}^{\infty} f_S(s)ds \qquad (7.14)$$

The probability for the stress value lying in the small interval ds with strength S exceeding the stress given by ds, under the assumption that the stress and strength random variables are independent, is given by

$$f_L(s)ds \int_{s_0}^{\infty} f_S(s)ds \qquad (7.15)$$

Hence, the reliability of an item is the probability that the strength S is greater than the stress L for all possible values of L, and can be determined by the following expression

$$R = P(S > L) = \int_{0}^{\infty} f_L(s) \left[\int_{s}^{\infty} f_S(s)ds \right] ds \qquad (7.16)$$

This is a general mathematical expression for the reliability of a family of products, or items whose strength is distributed when subject to a distributed load.

Reliability can also be computed on the basis that the stress remains less than the strength. Again assuming that the stress and the strength are independent random variables, the probability of the item for all possible values of the strength S is

$$R = P(S > L) = \int_0^\infty f_S(s) \left[\int_0^s f_L(s) ds \right] ds \qquad (7.17)$$

Carter (1972) incorporated the effects of time in the expression for reliability by introducing the parameter n which represents the number of times a load is repeatedly taken, each time from a distribution $f_L(s)$

$$R_n = \int_0^\infty f_S(s) \left[\left(\int_0^S f_L(s) ds \right)^n \right] ds \qquad (7.18)$$

Assuming that $n = \psi(t)$, reliability becomes a function of time. In general one can expect the function $\psi(t)$ to take a fairly complicated form, but for illustrative purposes, it is convenient to take a simple linear relation, and so

$$R_t = \int_0^\infty f_S(s) \left[\left(\int_0^S f_L(s) ds \right)^{Kt} \right] ds \qquad (7.19)$$

In conclusion, it can be said that the stress-strength presented approach neglects all time-dependent factors, but at the same time it is essential to point out that such an approach may well conceal the physical causes of failure. These causes may be purely random phenomena.

In order to illustrate the points made let us, for the moment, assume that the probability distribution of random variables S and L can be represented by the normal probability distribution. Consequently, the probability density function for a normally distributed strength S is given by

$$f_S(s) = \frac{1}{B_S \sqrt{2\pi}} \exp \left[-\frac{1}{2} \left(\frac{s - A_S}{B_S} \right)^2 \right] \qquad (7.20)$$

and the probability density function for a normally distributed stress L is

$$f_L(s) = \frac{1}{B_L \sqrt{2\pi}} \exp \left[-\frac{1}{2} \left(\frac{s - A_L}{B_L} \right)^2 \right] \qquad (7.21)$$

where A_S and B_S are scale and shape parameters of the strength, A_L and B_L are scale and shape parameters of the stress.

The probability distribution of the random variable, say G which represents the difference between random variables S and L, can also be described by the normal distribution with the following parameters

$$A_G = A_S - A_L$$

$$B_G = \sqrt{B_S^2 + B_L^2}$$

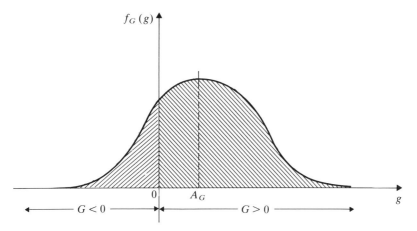

Figure 7.16 Reliability in cases of normal distributions for strength and stress.

According to Fig. 7.16 the reliability R can now be expressed in terms of G

$$R = P(G > 0) = \int_0^\infty \frac{1}{B_G \sqrt{2\pi}} \exp\left[-\frac{1}{2}\left(\frac{g - A_G}{B_G}\right)^2\right] dg \qquad (7.22)$$

If we let $z = (g - A_G)/B_G$, then $B_G dz = dG$. When $G = 0$, the lower limit of z is given by

$$z = \frac{0 - A_G}{B_G} = -\frac{A_S - A_L}{\sqrt{B_S^2 + B_L^2}} \qquad (7.23)$$

As strength and stress are normally distributed the random variable G is also normally distributed, and is fully defined by parameters A_G and B_G. When $g \to +\infty$, then the upper limit of $z \to +\infty$. Therefore

$$R = \frac{1}{\sqrt{2\pi}} \int_{\frac{A_S - A_L}{\sqrt{B_S^2 + B_L^2}}}^{\infty} \exp\left(\frac{-z^2}{2}\right) dz \qquad (7.24)$$

Clearly, the random variable Z is the standard normal variable. Consequently, the reliability can be found by the expression

$$R = 1 - \Phi\left(\frac{-A_S - A_L}{\sqrt{B_S^2 + B_L^2}}\right) \qquad (7.25)$$

Example 7.13 An item has been designed to withstand certain stresses. Based on experience it is expected that due to the variation in loading, the stress imposed on the item is normally distributed with a mean of 30 000 kPa and standard deviation of 5000 kPa. Due to variations in the material and dimensional tolerances it has been found that the strength is also normally distributed with a mean of 44 000 kPa and a standard deviation of 8000 kPa. Determine the reliability of the item under consideration.

SOLUTION According to the information available, the strength is defined as $N(44\,000, 8000)$ kPa, whereas the probability distribution of the load is fully defined as $N(30\,000, 5000)$ kPa. Making use of Eq. (7.23) the lower limit of the integral for R will be

$$z = -\frac{44000 - 30000}{\sqrt{(8000)^2 + (5000)^2}} = \frac{-14000}{9434} = -1.48$$

According to Table T1, $R = 0.9306$.

See Treson (1966), for derived expressions for reliability, in cases where different types of probability distributions are used to describe the distribution of stress and strength.

7.3 CONDITION PARAMETER APPROACH TO RELIABILITY

The approaches to reliability presented so far view the item as a black box which performs a required function until it fails. Such an approach is quite satisfactory for the statistician, but from an engineering viewpoint it is less so because engineers, especially operation and maintenance engineers, wish to understand the processes that are occurring inside the box. It is understood that the change to the condition of the item is a continuous process during its operating life and that an item is able to perform its required function as long as its condition remains within the acceptable range. In order to express the reliability of an item through its condition it is necessary to describe the condition at any instant of time.

7.3.1 Description of item condition

In order to describe the condition of an engineering item the concept of a condition parameter is introduced. A condition parameter can be any characteristic which is directly or indirectly connected with the item and its performance and which describes the condition of the item during its operating life. In every engineering item it is possible to detect several such characteristics, only some of which will satisfy the following requirements:

(a) Full description of the condition of an item
(b) Continuous and monotonic change during operating time
(c) Numerical definition of the condition of the item.

A condition parameter which satisfies all of these requirements is called the *relevant condition parameter, RCP*, because its numerical value fully describes and quantifies the condition of the item during operational life (Knezevic, 1983). Developing this approach an item is in a state of functioning as long as the relevant condition parameter lies between limits which are defined by the initial value, *RCPin*, and limiting value, *RCPlim*. When this parameter goes beyond the prescribed limits, the item or items begins to operate unsatisfactorily, thus

$$SoFu = RCPin < RCP < RCPlim$$
$$SoFa = RCP > RCPlim$$

Studies of change in condition processes (described by the relevant condition parameter) show that they are random processes because it is impossible to predetermine how they will develop. A particular process may therefore be expressed by a series of curves, as in Fig. 7.17, each having a given probability of occurrence, and hence describable only by probability theory. At a stated instant of time the condition of the item can be described through the relevant condition parameter which can have any value between the initial value and the maximum possible value, *RCPmax*, thus

$$RCPin < RCP, t < RCPmax \tag{7.26}$$

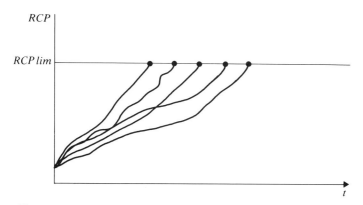

Figure 7.17 Change of condition during operation.

Therefore, at every instant of operating time, the relevant condition parameter, *RCP*, is a random variable which can only be expressed through its probability distribution. The probability density function of the relevant condition parameter at the stated time is denoted by $f_{RCP}(c, t)$, as shown in Fig. 7.18. Thus the probability of the value of the condition parameter being within the tolerance range for the stated time is also the probability of the reliable operation of the whole item

$$P(RCPin < RCP, t < RCPlim) = R(t) \tag{7.27}$$

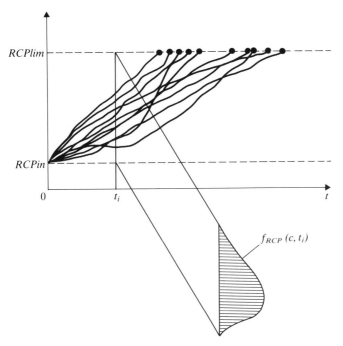

Figure 7.18 Condition parameter as a random variable.

The probability that RCP, at instant t, will have a value within the tolerance range, i.e. not exceeding the limiting value, can be defined as

$$P(RCPin < RCP, t < RCPlim) = \int_{RCPin}^{RCPlim} f_{RCP}(c, t) dc \qquad (7.28)$$

The above equation describes the probability that the random variable RCP at a stated time will have a value in the acceptable interval. However, in the case considered, the probability that the relevant condition parameter stays within the tolerance range at t, is also the probability that the item will remain functionable for a stated time, thus

$$R(t) = \int_{RCPin}^{RCPlim} f_{RCP}(c, t) dc \qquad (7.29)$$

The integral on the right of the equation represents the difference between cumulative distribution functions of the relevant condition parameter at instant t for the given limits. As the numerical value of the lower limit is obviously equal to zero, this can be rewritten

$$R(t) = F(c, t) \mid_{RCPin}^{RCPlim}$$

$$= F(RCPlim, t) \qquad (7.30)$$

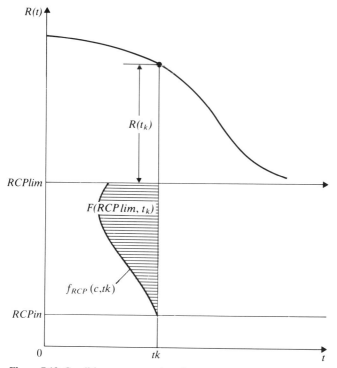

Figure 7.19 Condition parameter based approach to reliability.

A graphical description of the condition parameter based approach to reliability is presented in Fig. 7.19.

Example 7.14 In order to illustrate the method presented, let us assume that the relevant condition parameter, as a random variable at every instant of operating time, obeys the three-parameter Weibull distribution and that its initial value is 2 mm and limit value is 15 mm. At a stated operating time of 100 hours the parameters of the Weibull distribution have the following values: $A = 13.9$ and $B = 3.27$. Calculate the reliability at the stated time.

SOLUTION According to the task, $RCPin = 2$ and $RCPlim = 15$. Making use of Eq. (7.30), the reliability at stated time of 100 hours will be

$$R(100) = 1 - \exp\left[-\left(\frac{RCPlim - RCPin}{A - RCPin}\right)^B\right]$$

$$= 1 - \exp\left[-\left(\frac{15 - 2}{13.9 - 2}\right)^{3.27}\right] = 0.736902$$

A graphical interpretation of this example is presented in Fig. 7.20.

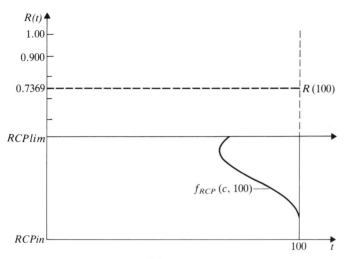

Figure 7.20 Reliability at stated time.

7.3.2 Condition-based approach to change in reliability

According to the condition-parameter-based approach to reliability presented above, it can be said that:

1. The reliability of the item can be expressed by the condition of the item.
2. The condition of the item at any instant of time can be described by the relevant condition parameter, *RCP*.
3. The state of failure is defined by the instant of time when the relevant condition parameter exceeds its limiting level, *RCP > RCPlim*.
4. The relevant condition parameter at any instant of time is a random variable, which is fully defined by its probability distribution function, e.g. $f_{RCP}(c, t)$.

In order to determine a way of describing the mechanism of change in condition let *RCP(t)* denote the random function of time which describes this random process. Changes in *RCP(t)* with the passage of time are conditioned both by external factors and by the course of physical processes that take place inside the item. One of the possible ways of expressing the probability distribution of continuous random variables is the probability density function, $f(.)$. The probability density function of the relevant condition parameter at the instant of operating time t was denoted by $f_{RCP}(c, t)$. There are an infinite number of such distributions, corresponding to the infinite number of possible instants of operating time. As a result accurate calculation of these distributions is associated with many insoluble mathematical problems.

Some simplifications are therefore necessary and we will assume that all probability functions belong to one family of probability distributions. This means that the distributive law of vertical intersection of random function $RCP(t)$ during operating time does not change. Even with this assumption it is impossible to determine parameters which define each of these distributions for every instant of time but it is possible to find numerical values for some of them. Knowing the probability distribution of the relevant condition parameter at several instances of time, $t_j, j = 1, m$, we still do not know the relationship between them throughout operating time.

In order to determine that relationship we will consider parameters which define these probability distributions, say $PT_i, i = 1, n$. In the case of normal and lognormal distributions $n = 2$, for Weibull $n = 2$ or 3, etc. Numerical values of these parameters can be approximated by a time-dependent function, say $PT_i(t) = \psi_i(t), i = 1, n$. These are called displacement functions. In order to determine them it is necessary to find as many points as possible which present the values of $PT_i(t_j)$ at different instances of operating time, $j = 1, m$. The displacement functions can be obtained quite simply by using a points diagram, on the basis of which a regression line can be drawn. The accuracy of results obtained for a description of the mechanism of change will directly depend on how close the assumptions are to reality.

This approach to reliability can be applied to the calculation of the mean time to failure, $MTTF$, which can be expressed by the equation

$$MTTF = \int_0^\infty F(RCPlim, t)dt \qquad (7.31)$$

and the reliability function represents the intersection between the random function $RCP(t)$ and the horizontal line defined as $RCP = RCPlim$, thus

$$R(t) = F(RCPlim, t) = \int_{RCPin}^{RCPlim} f_{RCP}(c, t)dc \qquad (7.32)$$

The approach presented is based on the condition of the item and its change during operating life. Such an approach provides a fuller picture of the condition of the item and its items throughout its lifetime because it is based on a continuous process of change rather than the time-to-failure approach which is based only on the moments of transition to a state of failure. Information about changes in condition is very valuable for engineers, particularly to the maintenance engineer who can base maintenance policy and strategy on this knowledge. This approach is generally applicable to all manufactured items, and especially those which are subjected to wear processes, which effect a gradual deterioration of material.

The main problems in the practical application of this approach are the selection of relevant condition parameters, and the determination of the parameters of probability distribution for $RCP(t)$. It cannot be taken for granted that a relevant condition parameter exists for every engineering item or that it will always be possible to find a function for the change of condition, due to the limitations of available equipment. This can limit the value of this approach.

Example 7.15 In order to illustrate this method for the description of the random function $RCP(t)$, let us assume that the relevant condition parameter, as a random variable for every instant of operating time, obeys the three-parameter Weibull distribution, with scale parameter A, shape parameter B and location parameter $C = RCPin$. This distribution is chosen in preference to other theoretical probability distributions because its range fully satisfies that of the relevant condition parameter, whereas the others go below $RCPin$ and may consequently introduce some numerical inaccuracies. The probability density function $f_{RCP}(c, t)$ will then take the following form

$$f_{RCP}(c, t) = \frac{B(t)}{A(t) - RCPin} \times \left(\frac{RCP - RCPin}{A(t) - RCPin} \right)^{B(t)-1}$$
$$\times \exp \left(-\frac{RCP - RCPin}{A(t) - RCPin} \right)^{B(t)} \tag{7.33}$$

Taking into account Eq. (7.3) the reliability function will be

$$R(t) = 1 - \exp \left(-\frac{RCPlim - RCPin}{A(t) - RCPin} \right)^{B(t)} \tag{7.34}$$

According to Eq. (7.5) the mean time to failure will have the following form

$$MTTF = \int_0^\infty \left[1 - \exp \left(-\frac{RCPlim - RCPin}{A(t) - RCPin} \right)^{B(t)} \right] dt \tag{7.35}$$

To illustrate the method, let us assume that the condition parameter obeys the three-parameter Weibull distribution, whose parameters are defined by the following time-dependent functions: scale parameter $A(t) = 0.1t + 5$, shape parameter $B(t) = -0.05t + 10$

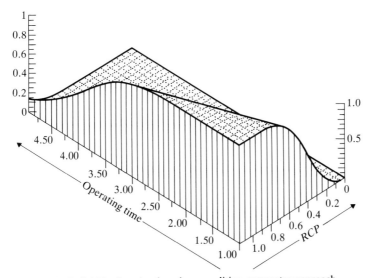

Figure 7.21 Reliability function based on condition parameter approach.

and minimum life parameter $C(t) = 0.1t + 2$. A graphical interpretation of an example where $RCPin = 2$ and $RCPlim = 15$ is presented in Fig. 7.21.

EIGHT

MEASURES OF MAINTAINABILITY

The main objective of this chapter is to define the measures through which maintainability can be quantitatively described and defined, just as functionability is numerically expressed through performance parameters. Earlier in the book it was demonstrated that the maintenance task represents a process which can only be described in probabilistic terms. Let us now establish a relationship between the concept of the probability system, given in Chapter 5 (see Fig. 5.14), and the concept of maintainability, presented in Chapter 3 (see Figs 3.2 and 3.3).

Restoration of the functionability of an engineering item can be considered as a random experiment and a system's transition to a state of functioning as an elementary event which represents its outcome. The function which assigns a corresponding numerical value t_i to every elementary event b_i from sample space S, is a random variable and is called *time to restore, TTR*, as shown in Fig. 8.1. Thus, the probability that the random variable *TTR* will take on value t_i is $p_i = P(TTR = t_i)$. The numerical values taken on by random variables, and the probability of their occurrence, define a probability distribution which can be expressed by different indicators. This establishes the complete analogy between the probability system defined in probability theory, and the ability of a system to be restored.

It should be noted at this point that although time is the true variable, it may often be more convenient to use some other readily available variables which represent time of use such as days, hours or minutes.

The aim of the above discussion is to introduce the concept of maintainability and show that it has a close relationship with probability, i.e. that by using the concept of the probability system, the maintainability of the system or items, as a qualitative characteristic, can be translated into a quantitative measure.

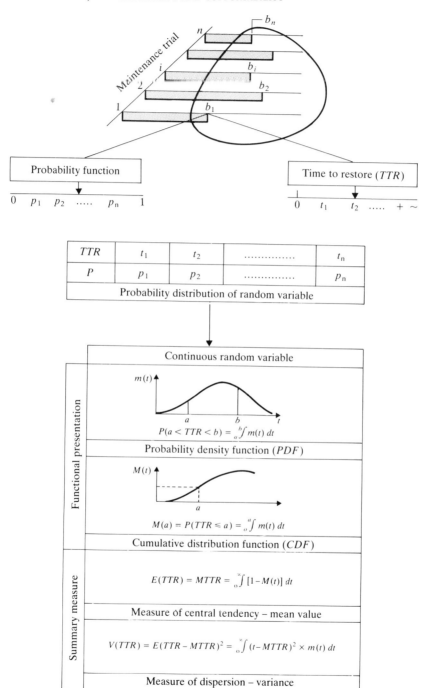

Figure 8.1 Concept of the probability system applied to the concept of maintainability.

8.1 MAINTAINABILITY CHARACTERISTICS

Since it is readily accepted that a population of supposedly identical copies of the item under consideration, restored under similar conditions, returns to *SoFu* at different instants of time, then it follows that a restoration process can only be described in probabilistic terms. Hence, maintainability is fully defined by the random variable *TTR* and its probability distribution, as shown in Fig. 8.1. The most frequently used characteristics of maintainability are:

1. Maintainability function
2. TTR_p time
3. Mean time to restore
4. Restoration success.

A brief definition and description of these characteristics follows.

8.1.1 Maintainability function

The cumulative distribution function of any random variable represents the probability that it will have a value equal to or less than some particular value, say *a*, $F(a) = P(X \le a)$. In the concept of maintainability the cumulative distribution function for the random variable *TTR* will be called the *maintainability function* and denoted by $M(t)$ (see Fig. 8.1). Thus, it presents the probability that the functionability of the system will be restored before or at the specified moment of maintenance time, *t*, thus

$$M(t) = P(\text{functionability will be restored before or at time } t)$$
$$= P(TTR \le t)$$
$$= \int_0^t m(t)dt \qquad (8.1)$$

where $m(t)$ is the probability density function of *TTR*, see Table 8.1.

It should be noted that in the case of the normal distribution the probability function exists from $-\infty$ so it may have a significant value at $t = 0$. Since negative time is meaningless in maintainability, great care should be used in manipulating this model unless $A > 3 \times B$, when the intercept at $t = 0$ can be considered negligible.

Example 8.1 The restoration time for a certain component can be represented by a normal distribution with parameters $A = 2$ and $B = 0.5$ hours. Plot the maintainability function and graphically determine the probability that the system will be restored within three hours.

SOLUTION According to Fig. 8.2, where the maintainability function is shown, the probability of restoring the system within the first three hours is 0.9772.

Table 8.1 Maintainability function, $M(t)$ for well-known theoretical distributions

Exponential	$1 - \exp(-t/A)$
	$t \geq 0$
Normal	$\Phi[(t - A)/B]$
	$-\infty \leq t \leq +\infty$
Lognormal	$\Phi[(\ln(t) - A)/B]$
	$t \geq 0$
Weibull	$1 - \exp -[(t - C)/(A - C)]^B$
	$t \geq 0$

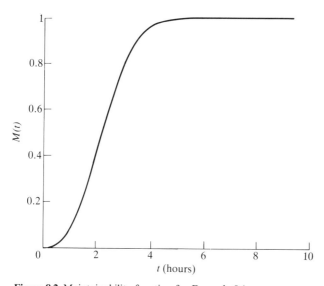

Figure 8.2 Maintainability function for Example 8.1.

8.1.2 TTR_p time

This is the maintenance time by which the functionability of a given percentage of a population will be restored. It is the abscissa of the point whose ordinate presents a given percentage of restoration. Mathematically, TTR_p time can be represented as

$$TTR_p = t \longrightarrow \text{ for which } M(t) = P(TTR \leq t) = \int_0^t m(t)dt = p \qquad (8.2)$$

The most frequently used is TTR_{90} time which presents the length of the restoration time by which 90 per cent of maintenance trials will be

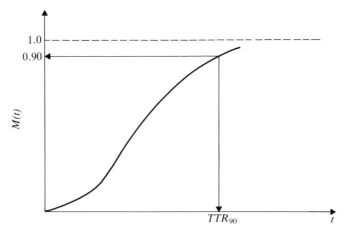

Figure 8.3 TTR_p time.

completed, as shown in Fig. 8.3.

$$TTR_{90} = t \longrightarrow \text{ for which } M(t) = P(TTR \leq s) = \int_0^t m(t)dt = 0.9$$

8.1.3 Expected restoration time

The expectation of the random variable TTR can be used for calculation of this characteristic of restoration process, thus

$$E(TTR) = \int_0^\infty t \times m(t)dt \qquad (8.3)$$

This characteristic is also known as the *mean time to restore* denoted as $MTTR$. Taking into account Eq. (5.21) the expected restoration time can be rewritten as

$$E(TTR) = MTTR = \int_0^\infty [1 - M(t)]dt \qquad (8.4)$$

which represents the area below the function which is complementary to the maintainability function. Table 8.2 shows expressions for the expected restoration time of well-known distributions.

Example 8.2 The functionability of a certain subsystem of the weather radar on a commercial aircraft can be restored by replacement of a specific component. Assume a Weibull distribution of the restoration time, $W(A=2.5$ hours, $B=2.7)$ and answer the following questions:

(a) What is the probability of restoring functionability within the first 180 minutes?
(b) What is the length of restoration time by which 99 per cent of the maintenance task will be completed?

Table 8.2 Expected Restoration
Time, $E(TTR)$, or $MTTR$, for
well-known distributions

Exponential	A
Normal	A
Lognormal	$\exp(A)$
Weibull	$A \times \Gamma(1 + 1/B)$

(c) What is the expected restoration time?

SOLUTION

(a)
$$P(TTR \leq 180) = F(3.0) = M(3.0) = \int_0^{3.0} m(t)dt$$

for the Weibull distribution it will be

$$M(3.0) = 1 - \exp(-(3.0/2.5)^{2.7}) = 0.8052$$

(b)
$$P(TTR \geq t) = 0.99 = 1 - \exp(-(t/2.5)^{2.7}), \quad \rightarrow t = 4.5 \text{ hours}$$

(c) According to Table 8.2 the expected restoration time can be calculated as follows

$$E(TTR) = MTTR = A \times \Gamma(1 + 1/(b)) = 2.5 \times 0.889 = 2.22 \text{ hours.}$$

8.1.4 Restoration success

The maintainability function, defined by Eq. (8.1), represents the probability
that the functionability of the item under consideration will be restored
before or at a specific instant of time t. In many real-life cases it is
important to know the probability that the item, which has not been
restored by time t_1, will be returned to *SoFu* by the time t_2. This problem
represents an example of conditional probability (see Sec. 5.2.4) because
the restoration could be completed by t_2 given that it was not at t_1. This
type of maintainability measure we shall call *restoration success*, $RS(t_1, t_2)$.
Making use of Eq. (5.9), this measure of maintainability is fully defined by
the expression

$$RS(t_1, t_2) = P(TTR < t_2 | TTR > t_1)$$

Making use of Eq. (8.1), which defines the maintainability function,
$M(t)$, and applying the principles of conditional probability (see Chapter 5),
the above expression for restoration success can be rewritten

$$RS(t_1, t_2) = P(TTR < t_2 | TTR > t_1) = \frac{M(t_2) - M(t_1)}{1 - M(t_1)} \quad (8.5)$$

In the case that the beginning of the interval coincides with the beginning of the restoration process, $t_1 = 0$, restoration success is equal to the

maintainability function at time t_2, thus

$$RS(t_1, t_2) = \frac{M(t_2) - M(t_1)}{1 - M(t_1)} = M(t_2)$$

since $M(0) = 0$. This measure of maintainability provides very useful information for maintenance engineers.

Example 8.3 The time to replace the clutch on a specific type of car can be represented by the Weibull distribution defined by parameters $A = 3$ hours and $B = 2.6$. Assuming that the restoration process has not been completed within the first three hours, determine the probability that the car will be returned to *SoFu* with a replaced clutch within the next hour.

SOLUTION According to the available data, the maintainability function is defined by

$$M(t) = 1 - \exp(-(t/3)^{2.6})$$

The probability that the maintenance task, which has not been completed by the end of the third hour of restoration work, will be completed within the following hour can be calculated in the following way

$$RS(3, 4) = [M(4) - M(3)]/[1 - M(3)] = 0.6668$$

where

$$M(3) = 1 - \exp(-(3/3)^{2.6}) = 0.632$$
$$M(4) = 1 - \exp(-(4/3)^{2.6}) = 0.8774$$

Thus, chances of completing the maintenance task by the end of the fourth hour, given that it has not been completed at the end of the third hour of restoration, is about 67 per cent.

NINE

MEASURES OF SUPPORTABILITY

The main objective of this chapter is to define the measures through which supportability can be quantitatively described and defined. It was demonstrated earlier in the book that the support task represents a process which can only be described in probabilistic terms. Let us now establish a relationship between the concept of the probability system, given in Chapter 5, and the concept of supportability, presented in Chapter 4.

The support task, whose main objective is provision of the support resources required for the performance of the specified maintenance task, can be considered as a random experiment and the task completion as an elementary event which represents its outcome. The function which assigns a corresponding numerical value t_i to every elementary event c_i from sample space S, is a random variable and will be called *time to support, TTS*, as shown in Fig. 9.1. Thus, the probability that the random variable TTS will take on value t_i is $p_i = P(TTS = t_i)$. The numerical values taken on by random variables, and the probability of their occurrence, define a probability distribution which can be expressed by different indicators. This establishes the complete analogy between the probability system defined in probability theory (see Fig. 5.14) and the ability of a system to be supported by specified resources during the restoration process (see Figs 4.2 and 4.3).

It should be noted at this point that although time is the true variable, it may often be more convenient to use some other readily available variables which represent time of use such as hours, days, weeks, or months.

This has introduced the concept of supportability and shown that it has a close relationship with probability, i.e. that by using probability, the supportability of an item as a qualitative characteristic can be translated into quantifiable measures.

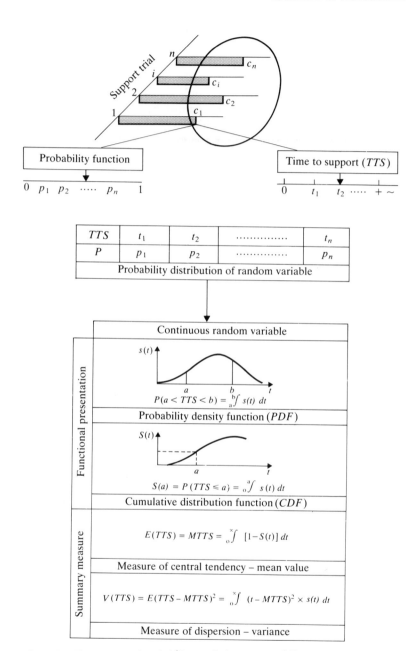

Figure 9.1 The concept of probability applied to supportability.

9.1 SUPPORTABILITY CHARACTERISTICS

Since it is readily accepted that a population of supposedly identical items experience states of failure for different lengths of time, it follows that the ability of the system to be supported can only be described in probabilistic terms. Hence, supportability is fully defined by the random variable *TTS* and its probability distribution, as illustrated by Fig. 9.1. The most frequently used characteristics of supportability are:

1. Supportability function
2. TTS_p time
3. Expected time to support
4. Support success.

A brief definition and description of these characteristics follows.

9.1.1 Supportability function

The cumulative distribution function of a random variable represents the probability that it will have a value equal to or less than some particular value, say a, $F(a) = P(X \leq a)$. In accordance with the concept of supportability, for the random variable *TTS*, the cumulative distribution function will be called the *supportability function* and denoted by $S(t)$ (see Fig. 9.1). At any instant of time t the supportability function presents the probability that the required support resources will be provided before or at the specified instant of time, t, thus

$$S(t) = P(\text{support resources will be provided before or at time } t)$$
$$= P(TTS \leq t)$$
$$= \int_0^t s(t)dt \qquad (9.1)$$

where $s(t)$ is the probability density function of *TTS*, see Table 9.1.

It should be noted that in the case of the normal distribution the probability function exists from $-\infty$ so it may have a significant value at $t = 0$. Since negative time is meaningless in supportability, great care should be used in manipulating this model unless $A > 3 \times B$, when the intercept at $t = 0$ can be considered negligible.

9.1.2 TTS_p time

This is the length of time by which required support resources will be provided, for a given percentage of demands. It is the abscissa of the point

Table 9.1 Supportability function, $S(t)$ for well-known theoretical distributions

Exponential	$1 - \exp{(-t/A)}$
	$t \geq 0$
Normal	$\Phi[(t - A)/B]$
	$-\infty \leq t \leq +\infty$
Lognormal	$\Phi[(\ln(t) - A)/B]$
	$t \geq 0$
Weibull	$1 - \exp{-[(t - C)/(A - C)]^B}$
	$t \geq 0$

whose ordinate presents a given percentage of the supportability function $S(t)$. TTS_p time can be mathematically represented as

$$TTS_p = t \rightarrow \text{ for which } S(t) = P(TTS < t) = \int_0^t s(t)dt = p \qquad (9.2)$$

The most frequently used is TTS_{90} which presents the length of time during which 90 per cent of support tasks will be completed, as shown in Fig. 9.2. This is expressed as:

$$TTS_{90} = t \rightarrow \text{ for which } S(t) = P(TTS < t) = \int_0^t s(t)dt = 0.90$$

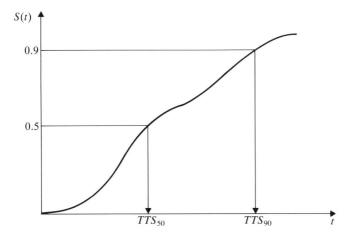

Figure 9.2 TTS_p time.

Table 9.2 Expected time to support, $E(TTS)$, or $MTTS$, for well-known distributions

Exponential	A
Normal	A
Lognormal	$\exp(A)$
Weibull	$A \times \Gamma(1 + 1/B)$

9.1.3 Expected time to support

The expectation of the random variable TTS can be used as another measure of supportability, thus

$$E(TTS) = \int_0^\infty ts(t)dt \qquad (9.3)$$

This characteristic is also known as *mean time to support, MTTS*. Making use of Eq. (5.21) this becomes

$$E(TTS) = MTTS = \int_0^\infty [1 - S(t)]dt \qquad (9.4)$$

Table 9.2 shows the expected time to support for the well-known probability distributions.

9.1.4 Support success

The supportability function, defined by Eq. (9.1), represents the probability that the support resources will be provided before or at a specific instant of time t. In many real-life cases it is important to know what is the probability that the support resources, which have not been provided by time t_1, will be provided by time t_2. This problem represents an example of conditional probability (see Sec. 5.2.4) because the resources can be provided by t_2 given that they were not available at t_1. This type of supportability measure we shall call *support success*, $SS(t_1, t_2)$. Making use of Eq. (5.9), this measure of supportability is fully defined by the following expression

$$SS(t_1, t_2) = P(TTS < t_2 | TTS > t_1)$$

Making use of Eq. (9.1) which defines the supportability function, $S(t)$, and applying the principles of conditional probability (see Chapter 5) the above expression for restoration success can be rewritten

$$SS(t_1, t_2) = P(TTS \leq t_2 | TTS > t_1) = \frac{S(t_2) - S(t_1)}{1 - S(t_1)} \qquad (9.5)$$

Thus, the support success is the conditional probability as defined by the above expression. In the case that the beginning of the interval coincides with the beginning of the support process, $t_1 = 0$, support success is equal to the supportability function at time t_2, thus:

$$SS\,(t_1, t_2) = \frac{S(t_2) - S(t_1)}{1 - S(t_1)} = S(t_2)$$

since $S(0) = 0$. This measure of supportability provides very useful information for operation and maintenance engineers.

Example 9.1 The functionability of an item in the weather radar equipment of a commercial aircraft can be restored by replacement of a specific component. Assume a Weibull distribution of the support time, $W(A{=}25$ hours, $B{=}2.7)$ and answer the following:

(a) What is the probability of providing support resources within the first 18 hours?
(b) What is the length of support time by which required resources will be provided in 90 per cent of cases?
(c) What is the expected support time?

SOLUTION

(a)
$$P(TTS \leq 18) = S(18) = \int_0^{18} s(t)\,dt$$

for the Weibull distribution it will be

$$S(18) = 1 - \exp(-(18/25)^{2.7}) = 0.3376$$

(b)
$$P(TTS \leq t) = 0.9 = 1 - \exp(-(t/25)^{2.7}), \quad \rightarrow t = 34 \text{ hours.}$$

(c) Referring to Table 9.2 the expected support time can be calculated as follows

$$E(TTS) = MTTS = A \times \Gamma(1 + 1/B) = 25 \times 0.889 = 22.2 \text{ hours}$$

where the numerical value for $\Gamma(1 + 1/2.7) = 0.889$ according to Table T4.

TEN

CONCEPT OF AVAILABILITY

In Chapter 1 it was concluded that any engineering item, from the point of view of functionability, at any instant of operating time, can be in one of two possible states: *SoFu* or *SoFa*. It is necessary to say that the transition from *SoFu* to *SoFa*, in cases of non-restorable items, happens only once during their operating life (see Fig. 1.1), whereas in the case of restorable items transitions between states are occurring consecutively throughout the operational lifetime, as shown in Fig. 1.2. Chapters 7, 8 and 9 showed that measures of reliability, maintainability and supportability are related to the time spent in *SoFu* and *SoFa*. Hence, the characteristic which quantitatively defines the relationship between them should be based on the corresponding variables.

Regarding the length of time spent in *SoFu* it could be said that this is determined by the characteristic known as reliability (Chapter 2). The entire time in *SoFa* is defined by maintainability, which is related to the time needed for restoration (Chapter 3), and supportability, which is related to the additional time incurred when the maintenance task cannot be performed due to lack of essential resources (Chapter 4).

The most important information for the user is to know the proportion of time for which the item is available for use. The measure which expresses this property is called the *availability*, *A*, and it is defined as:

$$A = \frac{\text{time in state of functioning}}{\text{time in state of functioning} + \text{time in state of failure}}$$

In order to express availability numerically, it is necessary to transform the above ratio into a mathematical relationship.

In Chapter 7 it was concluded that the time in the state of functioning is a random variable. Similarly, time spent in the state of failure can be expressed as random variables to describe the time to restore and the time to support, as discussed in Chapters 8 and 9. Thus, above expression for availability is a ratio between random variables which represents different random events. The most convenient way of describing the relevant random variables is by their expected values: $E(TTF) = MTTF$ for the time in *SoFu*, and $E(TTR) = MTTR$ and $E(TTS) = MTTS$ for the time in *SoFa*. Thus, the availability for engineering items can be expressed

$$A = \frac{E(TTF)}{E(TTF) + E(TTR) + E(TTS)} = \frac{MTTF}{MTTF + MTTR + MTTS} \quad (10.1)$$

This expression quantitatively represents the proportion of the operational life during which the item is available for use. It encompasses the inherent characteristics of design (reliability and maintainability), as well as the ability of the item to be supported by the user (supportability), and represents the probability that a restorable item will be in the state of functioning at any instant of time during its operational life.

Analysing Eq. (10.1) it is not difficult to conclude that its numerical value depends to a great extent on the way in which the user provides the specified support resources for the maintenance task. As a result, it is possible to obtain a distorted picture about the availability of the item as achieved by design. In order to obtain a more objective picture about the availability we shall neglect the value for $MTTS$. This type of availability is known as *inherent availability*, A_i and is defined as

$$A_i = \frac{E(TTF)}{E(TTF) + E(TTR)} = \frac{MTTF}{MTTF + MTTR} \quad (10.2)$$

The inherent availability represents the probability that an item will be in *SoFu* at any point in time while its functionability is restored, by performing specified maintenance activities with specified support resources. Inherent availability represents the achievement of the design, and it is greater or equal to the availability defined by Eq. (10.1). The difference between the two will depend on the facilities available to provide the required support resources, when needed.

Availability is a characteristic which quantitatively summarizes the functionability profile of an item. It is an extremely important and useful measure in cases where the user has to make decisions regarding the selection of an item from a wide choice of possibilities. For example, which item should the user select if X has the more favourable reliability measures, Y has superior maintainability, and Z can be supported the best? Clearly, this is a difficult task because the information given relates to different characteristics. Thus, in order to make an objective decision regarding the acquisition of the new item it is necessary to use information which encompasses all the related characteristics. Availability is the measure which provides this fuller picture about the functionability profile.

Example 10.1 The ability of a particular item in a car to maintain functionability during operating time can be represented by the Weibull distribution $W(A = 100, B = 4.3)$. The functionability of this item can be restored by replacement of a specific component. According to the designers, the restoration time can be represented by the Weibull distribution $W(A = 2.5, B = 2.7)$. According to the records obtained from a repair company, the length of time during which this type of car waits for the provision of a support resource can be described through the Weibull distribution $W(A = 4.2, B = 2.9)$. Assuming that the time in *SoFu* and *SoFa* are in hours, determine:

(a) The inherent availability of the considered subitem
(b) The availability of the subitem as achieved by the repair company.

SOLUTION
(a) In order to make use of Eq. (10.2) it is necessary to calculate mean time to failure of the item considered, as well as the mean time to restoration. These values are obtained in the following way

$$MTTF = A \times \Gamma(1 + 1/B) = 100 \times 0.910 = 91 \text{ hours}$$
$$MTTR = A \times \Gamma(1 + 1/B) = 2.5 \times 0.889 = 2.2 \text{ hours}$$

Thus, the achieved availability by design is

$$A_i = \frac{MTTF}{MTTF + MTTR} = \frac{91}{91 + 2.2} = 0.9763$$

(b) In order to determine the availability achieved by the repair company, it is necessary to calculate the mean support time, thus

$$E(TTS) = MTTS = A \times \Gamma(1 + 1/B) = 4.2 \times 0.909 = 3.8 \text{ hours}$$

Making use of Eq. (10.1), the availability of the item under consideration is

$$A = \frac{MTTF}{MTTF + MTTR + MTTS} = \frac{91}{91 + 2.2 + 3.8} = 0.9381$$

By comparison with the time achieved by design, it can be concluded that this particular company could increase the availability by improving those elements of support which significantly contribute to the length of time which the item spends in the state of failure.

ELEVEN

RELIABILITY PREDICTION

The main objective of this chapter is to analyse the engineering system as a set of items and to demonstrate the methodology for the prediction of reliability measures of the system based on the measures of individual consisting items (modules, units, components). This approach to the analysis of a system is known as the 'bottom-up' approach.

In Chapter 2 the concept of reliability was introduced, together with the measures which numerically describe the ability of the item to maintain functionability (Chapter 7). As previously mentioned, the word 'item' is used as a generic term which stands for: system, subsystem, module, component, part or unit.

11.1 RELIABILITY BLOCK DIAGRAM OF ENGINEERING SYSTEMS

In order to analyse the reliability of a system which consists of several items the concept of the reliability block diagram, RBD, is introduced. This is a diagrammatic representation of a system in which each consisting item is represented by a box. The relationship between items is determined by the effect that the failure of each one has on the functionability of the system as a whole. The structure of a block diagram for a particular system depends on its functionability definition. Developing the notation used in previous chapters the following new symbols are introduced:

- TTF_i—random variable for the *time to failure* of ith item
- TTF_s—random variable for the *time to failure* of the system.

The fundamental problem to be addressed here is how to predict the reliability measures of the system, given the reliability measures of the consisting items. The solution, catering for different configurations of the system, is given in the following sections.

11.1.1 Series configuration

• All consisting items must be functionable if the system is to be functionable.

The above definition fully describes the relationship between consisting items, and clearly states that the failure of any one of them causes the failure of the system as a whole. This type of configuration is probably the most commonly encountered in engineering practice. The reliability block diagram of a hypothetical system whose consisting items are connected in series is given in Fig. 11.1.

Figure 11.1 Reliability block diagram of a system with series configuration.

11.1.2 Parallel configuration

• Only one of the consisting items needs to be functionable for the system to be functionable.

The above definition clearly states that all consisting items must be in the state of failure for the system to fail. The reliability block diagram for a system whose consisting items are connected in parallel is shown in Fig. 11.2 which implies that:

(a) All items are in operation when the system is in operation
(b) Failures do not influence the reliability of the surviving items.

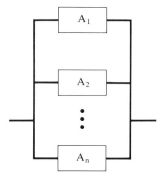

Figure 11.2 Reliability block diagram for parallel configuration.

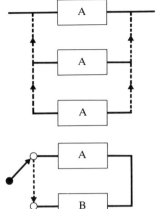

Figure 11.3 Standby parallel configuration.

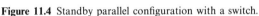

Figure 11.4 Standby parallel configuration with a switch.

It is particularly important to establish a parallel configuration in systems where reliability requirements are extremely high. The use of several engines in an aircraft is perhaps one of the most obvious examples. In spite of the fact that a parallel configuration increases reliability, it makes a system more complex, expensive and heavy, so that its application must be justified.

The type of parallel arrangement described above is known as a pure parallel configuration. In some cases it is possible to have a standby (off-load) parallel configuration. In this case some items are not brought into use until one operational item fails, Fig. 11.3. A good example of standby redundancy is the spare tyre in a motor vehicle. In some cases a switch exists to replace item A with item B when it is necessary, Fig. 11.4.

Another form of redundancy is an arrangement known as the r-out-of-n configuration. In this case the system has n items connected in parallel, but only r of them must be functionable at any instant of time for the system as a whole to be functionable. An example of this form of redundancy is the use of cables for lifts or suspension bridges, where only a certain number of them are necessary to support the weight, and the rest of them are employed for safety reasons (as standby), Fig. 11.5.

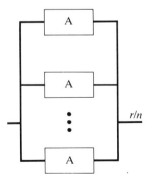

Figure 11.5 Standby redundant system, type r-out-of-n.

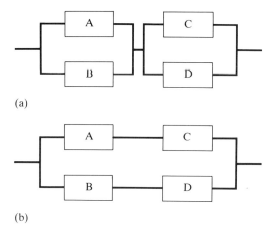

(a)

(b)

Figure 11.6 Parallel and series configurations of constituent items.

11.1.3 Complex configuration

In reality very few engineering systems have purely series or parallel configurations; most frequently they are a mixture of both. The combinations of parallel and series configurations are easily analysed by identifying and grouping consisting items into parallel or series configurations. As an example let us consider the configuration shown in Fig. 11.6(a).

To determine the reliability function of the system, it is necessary to group items connected in parallel, A and B first, and then C and D, in the series configuration. Now the system under consideration consists of two groups of consisting items which are connected in series. A second possible configuration is shown in Fig. 11.6(b). Here, two groups of items with a series configuration are connected in parallel.

The difference between the reliability of the system in these cases results from the different arrangement of consisting items.

11.2 PREDICTION OF SYSTEM RELIABILITY MEASURES

The most frequently used types of configuration have now been introduced, together with their RBD. The main objective of this section is to derive expressions for other reliability measures for the system based on:

(a) Reliability measures of consisting items
(b) Type of configuration.

The most important task in prediction is the determination of the overall reliability function of the system, because this fully determines all other measures.

11.2.1 Prediction of reliability function

Series configuration   The reliability function of a system whose consisting items are connected in series can be derived from the reliability functions of its consisting items. Thus, the reliability function of the system, $\mathcal{R}_s(t)$, which represents the probability that it will maintain functionability, with consisting items connected in series, is

$$\mathcal{R}_s(t) = P(TTF_s > t)$$
$$= P(TTF_1 > t \cap TTF_2 > t \cap TTF_3 > t \cap \text{.......} \cap TTF_{NCI} > t)$$

where NCI is the number of consisting items. This expression describes mathematically the system whose consisting items are connected in series as an intersection of events, whose probabilities are defined as $P(TTF_i > t)$, $i = 1, NCI$. It clearly states that the system under consideration will be functionable if, and only if, each consisting item is functionable.

In the case that random variables TTF_i, $i = 1, NCI$ represent independent events, the above expression becomes

$$\mathcal{R}_s(t) = P(TTF_1 > t) \times P(TTF_2 > t) \times \ldots \times P(TTF_n > t) = \prod_{i=1}^{n} P(TTF_i > t)$$

According to Eq. (7.3), the expression on the right-hand side is a reliability function of the ith item. Consequently, the reliability function of the system, which contains any number of items, $NCI > 2$, can be calculated

$$\mathcal{R}_s(t) = \prod_{i=1}^{NCI} R_i(t) \qquad (11.1)$$

The right-hand side, above, represents the product of the reliability functions of the consisting items and it is known as the product rule. It is necessary to emphasize that this expression can become very complex. The reason for this is that reliability functions of consisting items, $R_i(t)$ where $i = 1, NCI$ can be defined through any of the theoretical probability distributions and in the majority of cases their products cannot be expressed by any of the well-known distribution functions. For example, if two consisting items are connected in series and their reliability functions are expressed with, say, the normal or Weibull distribution, the reliability function of the system as a whole cannot be defined by either of these distributions. The situation becomes worse as the number of items increases. Therefore, Eq. (11.1) defines the reliability function of the system at discrete points in time as the product of the reliability functions of the consisting items at chosen instances of time. Likewise it does not define the probability distribution of the random variable TTF_s in single analytical form.

The only exception to the above statement are cases where the time to failure, of all the consisting items involved, can be described by the exponential distribution. The reason for this is the nature of the exponential distribution. The reliability function of a system whose consisting reliability functions are defined by the exponential distribution is determined by the expression

$$\mathcal{R}_s(t) = \prod_{i=1}^{NCI} \exp\left(-\frac{t}{A_i}\right)$$

$$= \exp\left(-\frac{t}{A_s}\right) \tag{11.2}$$

where

$$A_s = \frac{1}{\sum_{i=1}^{NCI} \frac{1}{A_i}}$$

Taking into account the relationship between the scale parameter A and failure rate λ in the case of the exponential distribution, $A = 1/\lambda$, the above expression can be rewritten

$$\mathcal{R}_s(t) = \prod_{i=1}^{n} \exp(-\lambda_i t)$$

$$= \exp(-\lambda_s t) \tag{11.3}$$

where

$$\lambda_s = \sum_{i=1}^{NCI} \lambda_i$$

Thus, the above expression defines analytically the probability distribution of the complex random variable TTF_s, which obeys the exponential distribution as well.

The simplicity of Eqs. (11.2) and (11.3) is the reason why the exponential distribution is the most frequently used in literature on reliability, for the analysis of complex systems. Again it is necessary to stress that these equations are valid *only* in cases where the reliability functions of all consisting items are defined through the exponential distribution.

Analysing Eq. (11.1) it can be deduced that the probability of the system being in *SoFu* at a specified moment of time will always be less than or equal to the lowest probability of all the consisting items, thus

$$\mathcal{R}_s(t) < \{R_i(t), \quad i = 1, NCI\}_{min}$$

Table 11.1 Reliabilities for various numbers of items

NCI in the system	$R_i(t)$	$\mathcal{R}_s(t)$
10	0.99	0.904
100	0.99	0.366
500	0.999	0.606
1000	0.999	0.367

Table 11.1 clearly illustrates these points. A system with 200 parts at 0.99 reliability each will have an overall reliability of 0.134, which means that the system under consideration has a chance of only 13 per cent of being functionable at the end of the stated time.

Example 11.1 A system consists of four items, A, B, C, and D, connected in series. The reliability measures of individual items are defined as follows: exponential distribution for item A, E(1000); Weibull distribution for item B, W(1200, 3.2); normal distribution for item C, N(800, 350); and Weibull distribution for item D, W(2000, 1.75). All parameters are in hours. Derive the expression for the reliability function of the system, $\mathcal{R}_s(t)$, and calculate the probability of the system being in *SoFu*, after the first 500, 750 and 1000 hours of use.

SOLUTION According to the information given and Table 7.2, the reliability functions of the items can be expressed as

$$R_A(t) = \exp\left(-\frac{t}{1000}\right)$$

$$R_B(t) = \exp\left[-\left(\frac{t}{1200}\right)^{3.2}\right]$$

$$R_C(t) = \int_t^\infty \frac{1}{350\sqrt{2\pi}} \exp\left[-\frac{1}{2}\left(\frac{t-800}{350}\right)^2\right] = \Phi\left(\frac{800-t}{350}\right)$$

$$R_D(t) = \exp\left[-\left(\frac{t}{2000}\right)^{1.75}\right]$$

As all items are connected in series (from the point of view of reliability), the reliability function of the system is defined by Eq. (11.1), thus

$$\mathcal{R}_s(t) = R_A(t) \times R_B(t) \times R_C(t) \times R_D(t)$$
$$= \exp\left(-\frac{t}{1000}\right)$$
$$\times \exp\left[-\left(\frac{t}{1200}\right)^{3.2}\right]$$
$$\times \int_t^\infty \frac{1}{350\sqrt{2\pi}} \exp\left[-\frac{1}{2}\left(\frac{t-800}{350}\right)^2\right]$$
$$\times \exp\left[-\left(\frac{t}{2000}\right)^{1.75}\right]$$

Table 11.2 Reliability functions for Example 11.1

t	$R_A(t)$	$R_B(t)$	$R_C(t)$	$R_D(t)$	$\mathscr{R}_s(t)$
500	0.60653	0.94109	0.80434	0.91541	0.42027
750	0.47237	0.80073	0.55686	0.83552	0.17598
1000	0.36788	0.57236	0.28382	0.74282	0.04439

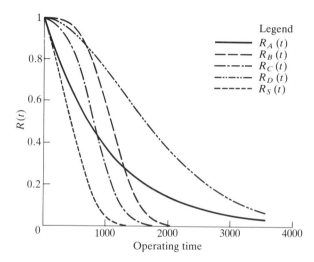

Figure 11.7 Reliability function of the system and its constituent items, Example 11.1.

Clearly this expression is not as easily used, but it does fully define the probability of the system being in *SoFu* at any instant of time, $0 < TTF_s < \infty$. In order to provide the answers to the questions asked, the random variable TTF_i will take specific values of 500, 750 and 1000. The values obtained for the reliability functions are shown in Table 11.2. The reliability function of the system will have a shape as shown in Fig. 11.7.

Example 11.2 Assume that the random variables which describe the time to failure of all four items given in Example 11.1 can be modelled by the exponential distribution function and they are defined as: item A, $E(1000)$; item B, $E(1200)$; item C, $E(800)$; and item D, $E(2000)$, where all distribution parameters are in hours. Derive the expression for the reliability function of the system, $\mathscr{R}_s(t)$, and calculate the probability of the system being in *SoFu* during the first 500, 750 and 1000 hours of operation.

SOLUTION According to the information given and Table 7.2, the reliability functions of the items can be expressed as:

$$R_A(t) = \exp\left(-\frac{t}{1000}\right)$$

$$R_B(t) = \exp\left(-\frac{t}{1200}\right)$$

$$R_C(t) = \exp\left(-\frac{t}{800}\right)$$

$$R_D(t) = \exp\left(-\frac{t}{2000}\right)$$

As all items are connected in series, from the point of view of reliability, the reliability function of the system is defined by Eq. (11.1), thus

$$\mathscr{R}_s(t) = R_A(t) \times R_B(t) \times R_C(t) \times R_D(t)$$

$$= \exp(-\frac{t}{1000}) \times \exp(-\frac{t}{1200}) \times \exp(-\frac{t}{800}) \times \exp(-\frac{t}{2000})$$

In this case it is possible to simplify the above expression for the reliability function of the system because the reliability functions of the consisting items are defined through the exponential distribution. Thus, making use of Eq. (11.2) the reliability function of the system, $\mathscr{R}_s(t)$ will have the following form

$$\mathscr{R}_s(t) = \exp(-\frac{t}{A_s}) = \exp(-\frac{t}{279.1})$$

In order to answer the questions, the random variable TTF_s will take specific values for which the corresponding reliability functions are as shown in Table 11.3. Figure 11.8 shows the reliability function of the system as well as reliability functions of the consisting items.

Table 11.3 Reliability functions for Example 11.2

A	1000	1200	800	2000	279.1
λ	0.001	0.000833	0.00125	0.0005	0.003583
t	$R_A(t)$	$R_B(t)$	$R_C(t)$	$R_D(t)$	$\mathscr{R}_s(t)$
500	0.60653	0.65924	0.53526	0.77880	0.16668
750	0.47237	0.53526	0.39161	0.68729	0.06805
1000	0.36788	0.43460	0.28650	0.60653	0.02778

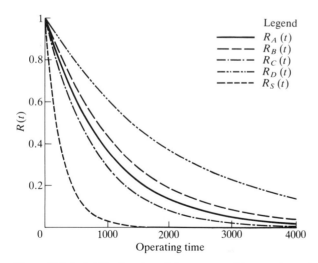

Figure 11.8 Reliability function of the system and its constituent items, Example 11.2.

Comparing the values for the reliability of the system shown in Tables 11.2 and 11.3 it is clear that the differences are significant and that the type of distribution function used for the description of the reliability functions plays a vital role.

Parallel configuration   The reliability function of the system whose consisting items are connected in parallel will be derived from the point of view of failure. If the failure function of the system is denoted as $\mathscr{F}_s(t) = P(TTF_s < t)$, the mathematical interpretation will have the following form

$$\mathscr{F}_s(t) = P(TTF_1 < t \cap TTF_2 < t \cap \ldots \cap TTF_{NCI} < t)$$

This clearly defines the probability of the event 'failure of the system' as the intersection of probabilities of the events 'failure of the consisting items'. Assuming individual events are independent, the above expression becomes

$$\begin{aligned}\mathscr{F}_s(t) &= P(TTF_s < t) \\ &= P(TTF_1 < t) \times P(TTF_2 < t) \times \ldots \times P(TTF_{NCI} < t) \\ &= \prod_{i=1}^{NCI} P(TTF_i < t)\end{aligned}$$

The probability of the expression on the right-hand side of the equation represents the cumulative distribution function of the random variable TTF_i, i.e. the failure function of ith item, which, according to Eq. (7.1), is denoted as $F_i(t) = P(TTF_i < t)$ thus

$$\begin{aligned}\mathscr{F}_s(t) &= P(TTF_s < t) \\ &= F_1(t) \times F_2(t) \times \ldots \times F_{NCI}(t) \\ &= \prod_{i=1}^{NCI} F_i(t)\end{aligned}$$

where $F_i(t)$ is a failure function of ith item. Making use of equation $F_i(t) + R_i(t) = 1$, the above expression can be written as

$$\mathscr{F}_s(t) = \prod_{i=1}^{NCI} (1 - R_i(t))$$

Finally, the expression for the reliability function of a system whose consisting items are connected in parallel is

$$\mathscr{R}_s(t) = 1 - \prod_{i=1}^{NCI} (1 - R_i(t)) \tag{11.4}$$

The above defines the reliability function for the system at discrete points. The same type of argument is applicable here as in the case of the series configuration of the system. It is necessary to say that in the case of parallel configurations it is extremely difficult to define the probability distribution function of the random variable TTF_s in a simple analytical form, even in cases where the reliability functions of the consisting items are defined by the exponential distribution. Application of a parallel configuration, for the consisting items, greatly increases reliability of the system. This means that it is possible to increase the probability of maintaining the functionability of systems while using consisting items with lower reliability

$$\mathcal{R}_s(t) > \{R_i(t), i = 1, NCI\}_{\max}$$

Example 11.3 A system consists of four items, A, B, C, and D, connected in parallel. The reliability functions of the individual items are: item A, E(1000); item B, W(1200, 3.2); item C, N(800, 350); item D, W(2000, 1.75). All parameters are in hours. Derive the expression for the reliability function of the system, $\mathcal{R}_s(t)$, and calculate the probability of the system being in SoFu, after the first 500, 750 and 1000 hours of use.

SOLUTION According to the information given and Table 7.2, the reliability functions of the items can be expressed as

$$R_A(t) = \exp\left(-\frac{t}{1000}\right)$$

$$R_B(t) = \exp\left[-\left(\frac{t}{1200}\right)^{3.2}\right]$$

$$R_C(t) = \int_t^\infty \frac{1}{350\sqrt{2\pi}} \exp\left[-\frac{1}{2}\left(\frac{t-800}{350}\right)^2\right] = \Phi\left(\frac{800-t}{350}\right)$$

$$R_D(t) = \exp\left[-\left(\frac{t}{2000}\right)^{1.75}\right]$$

As all items are connected in parallel the reliability function of the system is defined by Eq. (11.4), thus

$$\mathcal{R}_s(t) = 1 - [(1 - R_A(t)) \times (1 - R_B(t)) \times (1 - R_C(t)) \times (1 - R_D(t))]$$

$$= 1 - (1 - \exp(-\frac{t}{1000}))$$

$$\times (1 - \exp\left[-\left(\frac{t}{1200}\right)^{3.2}\right])$$

$$\times (1 - \int_t^\infty \frac{1}{350\sqrt{2\pi}} \exp\left[-\frac{1}{2}\left(\frac{t-800}{350}\right)^2\right])$$

$$\times (1 - \exp\left[-\left(\frac{t}{2000}\right)^{1.75}\right])$$

Clearly this expression is not easy to use, but it fully defines the probability of the system being in SoFu at any instant of time, $0 < TTF < \infty$. In order to answer the questions asked, the random variable TTF_s will take specific values whose corresponding reliability functions are shown in Table 11.4. The reliability function of the system and its consisting items, which are connected in parallel, are shown in Fig. 11.9.

The influence of configuration type is clearly demonstrated by the numerical values for the reliability functions shown in Tables 11.2 and 11.4.

Table 11.4 Reliability functions for Example 11.3

t	$1-R_A(t)$	$1-R_B(t)$	$1-R_C(t)$	$1-R_D(t)$	$\mathcal{R}_s(t)$
500	0.39347	0.05892	0.19566	0.08459	0.999616
750	0.52763	0.19927	0.44314	0.16448	0.992336
1000	0.63212	0.42744	0.71618	0.25718	0.950211

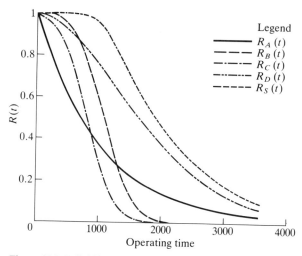

Figure 11.9 Reliability function of the system and its constituent items, Example 11.3.

Parallel configuration type _r_-out-of-_n_   The reliability function for the system whose consisting items are connected in parallel with the special type known as _r_-out-of-_n_ is defined by the expression

$$\mathcal{R}_s(t) = \sum_{x=r}^{NCI} (C_x^{NCI} R_I^x(t) \times (1 - R_I(t))^{NCI-x}) \tag{11.5}$$

where $R_I(t)$ is the item reliability function, assumed to be equal for all items, and C_x^{NCI} is the total number of combinations of NCI consisting items, taken x at a time.

Example 11.4 For a system to work three modules have to be functionable. In order to increase the probability that the system will maintain functionability, it has been decided to incorporate five modules. Derive the expression for the reliability function of the system assuming that the reliability function for the modules is defined by the Weibull distribution, $W(2000, 3.5)$, and calculate the probability of the system being in _SoFu_ after the first 1200 hours of service.

SOLUTION The reliability function for each module is defined by the following expression

$$R_I(t) = R_m(t) = \exp(-(\frac{t}{2000})^{3.5})$$

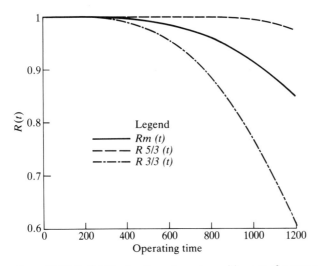

Figure 11.10 Reliability function of a system with r-out-of-n type configuration.

According to the chosen configuration, the system is functionable if three out of five modules are functionable. Thus, the reliability function for the system is defined by Eq. (11.5), where $n = 5$, and $r = 3$.

$$\mathcal{R}_s = \sum_{x=3}^{5} (C_x^5 \times (R_m^x(t) \times (1 - R_m(t))^{5-x}) =$$
$$= 5!/(3!(5-3)!) \times (R_m^3(t) \times (1 - R_m(t))^2)$$
$$+ 5!/(4!(5-4)!) \times (R_m^4(t) \times (1 - R_m(t))^1)$$
$$+ 5!/(5!(5-5)!) \times (R_m^5(t) \times (1 - R_m(t))^0)$$

This expression fully defines the reliability function of the system with a parallel configuration of type 3-out-of-5. The probability that the system will be functionable after the first 1200 hours of operation can be calculated by making use of the above function. Thus

$$R_m(1200) = \exp(-(1200/2000)^{3.5}) = 0.8459$$
$$\mathcal{R}_s(1200) = 0.1382 + 0.3915 + 0.4437 = 0.9734$$

Consequently it can be seen that the probability that the system will maintain functionability during the first 1200 hours is considerably increased by the use of two additional modules ($0.9734 \gg 0.6141$), as shown in Fig. 11.10.

Complex configuration As an example let us consider a system whose consisting items are connected in series and in parallel as shown in Fig. 11.11(a).

To determine the reliability function of the system, it is necessary to group items connected in parallel, A and B first, and then C and D, in the series configuration. Thus, the reliability function for A and B will be $R_{AB}(t) = 1 - [(1 - R_A(t))(1 - R_B(t))]$ and this is applicable to C and D so that $R_{CD}(t) = 1 - [(1 - R_C(t))(1 - R_D(t))]$. The system now consists of two

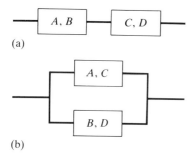

(a)

(b)

          Figure 11.11 Parallel and series configuration.

groups of items, connected in series. Using Eq. (11.1) the reliability function for the system will be

$$\mathcal{R}_s(t) = R_{AB}(t) \times R_{CD}(t)$$
$$= [1 - (1 - R_A(t))(1 - R_B(t))] \times [1 - (1 - R_C(t))(1 - R_D(t))]$$

A second possible configuration for the same items is shown in Fig. 11.11 (b). Here, two groups of items with series configuration are connected in parallel. The procedure to determine the reliability function of the system is similar to the first case. Using Eq. (11.1) the reliability functions of the series configuration will be: $R_{AC}(t) = R_A(t) \times R_C(t)$, and $R_{BD}(t) = R_B(t) \times R_D(t)$.

Combining the above expressions using Eq. (11.4) for the parallel configuration, the reliability function of the system takes the following form

$$\mathcal{R}_s(t) = 1 - [(1 - R_{AC}(t))(1 - R_{BD}(t))]$$
$$= 1 - [(1 - R_A(t)R_C(t))(1 - R_B(t)R_D(t))]$$

The difference between the reliability of the system in these two cases results from the different arrangement of consisting items.

11.2.2 Prediction of the failure function

As a system at any instant of time can be in one of two possible states, the failure function of the system as a whole, $\mathcal{F}_s(t)$, can be expressed

$$\mathcal{F}_s(t) = 1 - \mathcal{R}_s(t) \tag{11.6}$$

as shown in Fig. 7.3. Thus, the above expression is valid for any system, regardless of its configuration type.

Example 11.5 A system consists of five items with the following $MTBFs$: item A, 10 540 hrs; item B, 16 220 hrs; item C, 9500 hrs; item D, 12 100 hrs and item E, 3600 hrs. Assuming that all are connected in series, determine the probability of the system failing to maintain functionability during the first 1000 hours of operation.

SOLUTION According to the data provided, all necessary information is shown in Table 11.5. Based on this information the failure rate of the system is 0.0006206. Making use of Eqs (11.2) and (11.3) and using the data from Table 11.5 the reliability function can

Table 11.5 Example 11.5 data

item	A	B	C	D	E
$MTTF$	10 540	16 220	9500	12 100	3600
λ	0.000095	0.0000616	0.000105	0.000082	0.000277

be calculated according to the following expression

$$\mathscr{R}_s(1000) = \exp(-0.00062 \times 1000) = 0.5379$$

The required probability can be directly calculated by using Eq. (11.6), thus

$$\mathscr{F}_s(1000) = 1 - \mathscr{R}_s(1000) = 0.4621$$

11.2.3 Prediction of TTF_p time

This is the length of operational time by which the functionability of a given percentage of a population will have failed. It is the abscissa of the point whose ordinate presents a given percentage of failure. Mathematically, TTF_p time can be represented as

$$TTF_p = t \longrightarrow \text{ for which } \mathscr{F}_s(t) = P(TTF \le t) = \int_0^t f(t)dt = p \qquad (11.7)$$

Example 11.6 For the data given in Example 11.1, determine the operational time up to which 85 per cent of systems will have failed.

SOLUTION According to Eq. (11.7) the task is to determine the value of t for which $\mathscr{F}_s(t) = 0.85$, thus

$$TTF_{85} = t \longrightarrow \text{ for which } \mathscr{F}_s(t) = P(TTF \le t) = \int_0^t f(t)dt = 0.85$$

Looking at this very complex expression it appears that the solution to the problem can be more easily found by using a graphical presentation of this function. According to Figs 11.7 and 11.12 the operational time by which 85 per cent of operational systems will fail is about 810 hours.

11.2.4 Prediction of expected time to failure

Making use of Eq. (7.5), the expression for the expected operating time of the system, $E(TTF_s)$, or the mean time to failure for the system, $MTTF_s$, can be derived as

$$E(TTF_s) = MTTF_s = \int_0^{\infty} \mathscr{R}_s(t)dt \qquad (11.8)$$

Generally speaking, the easiest way of calculating the numerical value of $MTTF_s$ is by numerical integration. In cases where random variables TTF_i

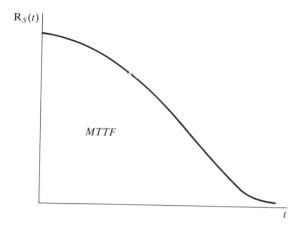

Figure 11.12 Reliability function for Example 11.6.

are defined through the exponential distribution, the mean time to failure of the system can be calculated according to the expression

$$MTTF_s = \frac{1}{\sum_{i=1}^{NCI} \frac{1}{MTTF_i}} \tag{11.9}$$

This expression is valid in cases where the consisting items are connected in series.

Example 11.7 A system consists of five items with the following $MTBF$s: item A, 10 540 hrs; item B, 16 220 hrs; item C, 9500 hrs; item D, 12 100 hrs and item E, 3600 hrs. The five items are connected in series. Determine the mean time to failure for the system as a whole.

SOLUTION The numerical value of $MTTF_s$ can be calculated by making use of Eq. (11.9), thus

$$MTTF_s = \frac{1}{\frac{1}{10\,540} + \frac{1}{16\,220} + \frac{1}{9500} + \frac{1}{12\,100} + \frac{1}{3600}} = 1611.34$$

Hence, the mean time to failure, or expected value of the random variable TTF_s, is 1611 hours.

11.2.5 Prediction of mission success

The probability that a system under consideration whose reliability function is derived as $\mathcal{R}_s(t)$ will be functionable at the end of a mission which begins at instant of time t_b can be determined by applying Eq. (7.6), thus

$$\mathcal{MS}_s(t_b, \text{mission}) = P(TTF_s > t_b + \text{mission}|TTF_s > t_b)$$
$$= \frac{\mathcal{R}_s(t_b + \text{mission})}{\mathcal{R}_s(t_b)} \tag{11.10}$$

Example 11.8 For the system analysed in Example 11.1, determine the probability of 250 and 500 hours mission success assuming that it has already been in operation for 500 hours.

SOLUTION Numerical values for the reliability function of the system are given in Table 11.2. Using Eq. (11.10), the mission success will be

$$\mathcal{MS}_s(500, 250) = \frac{\mathcal{R}_s(750)}{\mathcal{R}_s(500)} = \frac{0.1759}{0.42027} = 0.4185$$

$$\mathcal{MS}_s(500, 500) = \frac{\mathcal{R}_s(1000)}{\mathcal{R}_s(500)} = \frac{0.044392}{0.42027} = 0.1056$$

Example 11.9 A system consists of four items, A, B, C, and D, connected in series. The probability of items, A, B, C and D, being in $SoFu$ at the end of the mission is: $A = 0.97$, $B = 0.85$, $C = 0.90$ and $D = 0.85$. Determine the probability of the system being in $SoFu$ at the end of the planned mission, assuming that the beginning of the mission coincides with the beginning of the operation of the system.

SOLUTION According to Eq. (11.1), the probability that a serially configured system will be in $SoFu$ at a stated instant of time, Tst, is equal to the product of the probabilities that the individual items will be in $SoFu$. Assuming that the stated time, $Tst = 0 + $ mission, the required probability can be determined in the following way

$$\mathcal{R}_s(Tst) = R_A(Tst) \times R_B(Tst) \times R_C(Tst) \times R_D(Tst) = 0.97 \times 0.85 \times 0.90 \times 0.85 = 0.63$$

Making use of Eq. (11.10), the mission success will be

$$\mathcal{MS}_s(0, \text{mission}) = P(TTF_s > \text{mission}|TTF_s > 0)$$
$$= \frac{\mathcal{R}_s(\text{mission})}{\mathcal{R}_s(0)}$$
$$= 0.63/1.0 = 0.63$$

Example 11.10 For the successful completion of a mission, four vehicles have to be in $SoFu$ at the end of a planned journey. In order to increase the chance of mission completion it has been decided to send one extra vehicle. Determine the probability of mission completion assuming that each vehicle has a chance of success of 0.85.

SOLUTION The probability of mission completion for each motor vehicle is the reliability of the vehicle considered, thus: $R_{mv}(t) = 0.85$. In order to complete the mission successfully four out of five vehicles have to be in the state of functioning, so that $n = 5$, $r = 4$. The probability of mission completion R_m can be determined by applying Eq. (11.3)

$$R_m = \sum_{x=4}^{5} (C_x^5 \times (R_{mv}^x(\text{mission}) \times (1 - R_{mv}(\text{mission}))^{5-x})) =$$
$$= 5!/(4!(5-4)!) \times (0.85^4 \times (1-0.85)^1)$$
$$+ 5!/(5!(5-5)!) \times (0.85^5 \times (1-0.85)^0)$$
$$= 0.3915 + 0.4437 = 0.8352$$

Thus, the probability of mission completion is considerably increased by sending an extra vehicle ($0.8352 \gg 0.5222$).

11.2.6 Prediction of hazard function

According to Eq. (7.8) the hazard function of the item under consideration represents the ratio between the probability density function and the corresponding reliability function at a specified instant of time, $z(t) = f(t)/R(t)$. Unfortunately this expression cannot be used for the prediction of the hazard function of the system because, in the majority of cases, the function which defines the probability distribution of the random variable TTF_s is unknown, i.e. it is impossible to derive the analytical expression for it. Thus, in order to determine an expression for the prediction of the hazard function, regardless of the configuration of its consisting items and the type of theoretical probability distributions which defines their TTFs, Eq. (7.7) for the failure rate will be used. This is acceptable because, in Chapter 7, the hazard function was defined as a limit of the failure rate as the interval approaches zero. Thus, the hazard function for the system can be determined

$$z_s(t) = \frac{\mathscr{R}_s(t_1) - \mathscr{R}_s(t_2)}{(t_2 - t_1) \times \mathscr{R}_s(t_1)} \tag{11.11}$$

The above equation is applicable to any system and any type of configuration. Clearly, the accuracy of the above expression increases as the interval $t_2 - t_1$ decreases (see Eq. (7.8)).

Example 11.11 Let us look at a system which consists of three items A, B, and C, connected in series from the point of view of reliability. The parameters which define their probability distributions for time to failure are given in hours in Table 11.6. Determine the mean time to failure of the system and plot its hazard function.

SOLUTION The solution to this problem can be easily found by applying Eq. (11.9). Thus, the failure rate of each module is $1/1000=0.001$ and, as the modules are connected in series, the sum of their failure rates will give the failure rate of the system. Finally, the failure rate of the system is 0.003, which means that the mean time to failure of the system will be 333.3 hours.

A graphical representation of the hazard function is given by Fig. 11.13, and it is a horizontal line because of the constant failure rate of 0.003 failures per hour.

Let us resolve the same problem again, but this time instead of a constant failure rate we shall specify the probability distributions of the time to failure for the corresponding items, as shown in Table 11.7.

Table 11.6 Probability characteristics for Example 11.11

Item	Type of distribution	Parameter scale	shape	MTTF (hours)
A	exponential	1000	-	1000
B	exponential	1000	-	1000
C	exponential	1000	-	1000

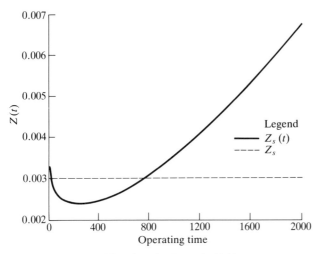

Figure 11.13 Hazard functions for Example 11.11.

Table 11.7 Probability characteristics for Example 11.11, specified distributions

Module	Type of distribution	Parameter scale	shape	MTTF (hours)	failure rate
A	normal	1000	475	1000	increasing
B	exponential	1000	-	1000	constant
C	Weibull	880	0.8	1000	decreasing

The mean time to failure of the system is defined by Eq. (11.8). Using a graphical method to determine its numerical value, the mean time to failure for the system is 362.2 hours. This represents an increase of 10 per cent in comparison with the corresponding value obtained originally. The differences between these two sets of data are even more significant when the hazard functions are compared, as illustrated by Fig. 11.13. The hazard function plotted has been obtained by making use of Eq. (11.11).

Example 11.12 The reliability block diagram for a hypothetical engineering system is shown in Fig. 11.14. The type of distribution function which defines the time to failure for each of the consisting items is given in Table 11.8, together with their parameters. Derive the reliability function of the system given, and determine:

(a) The probability that the system will be functionable after the first 300 hours of operation.
(b) The probability that the transition to *SoFa* will take place within the first 300 hours.
(c) The probability that the system will successfully complete a 100-hour mission which is expected to start after the first 300 hours.

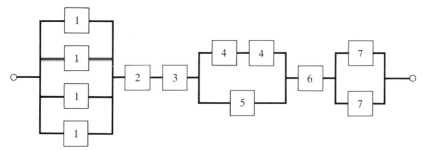

Figure 11.14 Reliability block diagram for Example 11.12.

Table 11.8 Probability distributions for
Example 11.12

Item	Type of distribution	Scale	Shape
1	Weibull	1000	1.7
2	exponential	2450	-
3	Weibull	1710	3.7
4	Weibull	3100	2.4
5	normal	980	310
6	normal	1800	650
7	exponential	400	-

SOLUTION Before we start analysing the reliability of the system we shall define the
reliability function of each consisting item. According to Table 7.2, the reliability functions
for consisting items are defined as

$$R_1(t) = \exp(-(\frac{t}{1000})^{1.7})$$

$$R_2(t) = \exp(-\frac{t}{2450})$$

$$R_3(t) = \exp(-(\frac{t}{1710})^{3.7})$$

$$R_4(t) = \exp(-(\frac{t}{3100})^{2.4})$$

$$R_5(t) = \int_{310}^{\infty} \frac{1}{310\sqrt{2\pi}} \exp(-(\frac{t-980}{310})^2) = \Phi(\frac{980-t}{310})$$

$$R_6(t) = \int_{850}^{\infty} \frac{1}{650\sqrt{2\pi}} \exp(-(\frac{t-1800}{650})^2) = \Phi(\frac{1800-t}{650})$$

$$R_7(t) = \exp(-\frac{t}{4000})$$

Clearly we are dealing with a complex configuration. In order to derive the reliability
function for the system it is necessary to go through several steps, each of which will
simplify the reliability block diagram. In the first step we shall group the consisting items

Table 11.9 Items grouped into modules, Example 11.12

Module	Type of configuration	Item
A	r out of n (3/4)	1
B	series	2,3
C	parallel	4,5
D	series	6
E	parallel	7

into modules, as shown in Table 11.9, and it is now clear that all modules are connected in series.

The reliability function for the system, $\mathcal{R}_s(t)$ will have the following form

$$\mathcal{R}_s(t) = R_A(t) \times R_B(t) \times R_C(t) \times R_D(t) \times R_E(t)$$

where the reliability functions of the modules are

$$R_A(t) = \sum_{x=3}^{4}(C_x^4 \times (R_1^x(t) \times (1 - R_1(t))^{4-x})$$
$$= 4!/(3!(4-3)!) \times (R_1^3(t) \times (1 - R_1(t))^1)$$
$$+ 4!/(4!(4-4)!) \times (R_1^4(t) \times (1 - R_1(t))^0)$$
$$= 4R_1^3(t) - 3R_1^4(t)$$
$$R_B(t) = R_2(t) \times R_3(t) = R_2(t)R_3(t)$$
$$R_C(t) = 1 - [(1 - R_4^2(t))(1 - R_5(t))]$$
$$R_D(t) = R_6(t)$$
$$R_E(t) = 1 - [1 - R_7(t)]^2$$

Finally, the reliability function can be derived by replacing the expressions for $R_i(t), i = 1, NCI$ into expressions for the reliability functions of the modules and by applying the product rule to them. Thus, the reliability function, for the system under consideration, will have the following form

$$\mathcal{R}_s(t) = 4[\exp(-(\frac{t}{1000})^{1.7})]^3 - 3[\exp(-(\frac{t}{1000})^{1.7})]^4$$
$$\times [\exp(-\frac{t}{2450}) \times \exp(-(\frac{t}{1710})^{3.7})]$$
$$\times [1 - (1 - \exp(-(\frac{t}{3100})^{2.4})^2)(1 - \Phi(\frac{980 - t}{310}))]$$
$$\times \Phi(\frac{1800 - t}{650})$$
$$\times 1 - [1 - \exp(-\frac{t}{4000})]^2$$

Clearly, the above expression is not simple, but further simplification is not possible. Thus it is impossible to define the probability distribution function for TTF_s expressed through a single well-known theoretical distribution.

(a) $\qquad \mathcal{R}_s(300) = 0.9155 \times 0.8833 \times 0.99989 \times 0.98949 \times 0.7216 = 0.58366$

(b) $\qquad \mathcal{F}_s(300) = 1 - \mathcal{R}_s(300) = 1 - 0.58366 = 0.41134$

(c) Making use of Eq. (11.10) the required probability for the mission success will be:

$$\mathcal{MSP}_s(300, 100) = \frac{\mathcal{R}_s(300 + 100)}{\mathcal{R}_s(300)}$$

and

$$\mathcal{R}_s(300 + 100) = 0.83436 \times 0.84544 \times 0.99955 \times 0.98437 \times 0.600423 = 0.416734$$

Thus, the probability of mission success is

$$\mathcal{MSP}_s(300, 100) = 0.416734/0.58366 = 0.714$$

TWELVE

MAINTAINABILITY PREDICTION

This chapter is intended to analyse a complex maintenance task as a set of independent maintenance tasks, and demonstrate a methodology for deriving expressions which predict maintainability measures, based on the corresponding measures for the consisting tasks.

In Chapter 3 the concept of maintainability was introduced. In Chapter 8 the measures which numerically describe the ability of the item to be restored, by performing a specified maintenance task, were discussed. As previously mentioned the maintenance task represents a set of maintenance activities which are required in order to restore the functionability of the item under consideration.

12.1 COMPLEX MAINTENANCE TASK

A complex maintenance task is defined as:

A set of simultaneously performed maintenance tasks all of which must be completed if the task is to be completed.

This definition fully describes the relationship between comprising tasks and clearly states that:

(a) All tasks start at the same instant of time and they are performed simultaneously.
(b) Failure to complete any of the consisting tasks causes the failure of the task as a whole.

A typical example of a complex maintenance task is the 6000 (10 000) miles service for motor vehicles, as required by their manufacturers. The complex task consists of tasks which are related to the engine, transmission, brakes, starting system, electrical items and so forth. The whole task is complete only when all of the consisting tasks have been accomplished.

Developing the notation used in Chapter 8 the following symbols are introduced:

- $TTCT_i$—random variable for the time to complete the ith maintenance task.
- TTR—random variable for the time to restore the functionability of the system (or module) considered, by completing specified maintenance tasks.

The fundamental question here is how to predict maintainability measures for a complex maintenance task, given that maintainability measures for the consisting tasks are known.

12.2 PREDICTION OF MAINTAINABILITY MEASURES FOR COMPLEX TASKS

Maintainability measures of a complex task whose consisting tasks are performed simultaneously can be derived from the maintainability measures of the individual tasks. Thus, the maintainability function for the overall maintenance task, $\mathcal{M}_t(t)$, which represents the probability that functionability will be restored by the certain instant of time, t, is defined as

$$\mathcal{M}(t) = P(TTR_t < t)$$
$$= P(TTCT_1 < t \cap TTCT_2 < t \cap TTCT_3 < t \cap \ldots \cap TTCT_{NCT} < t)$$

where NCT is the number of consisting tasks.

The above expression mathematically describes the complex maintenance task as an intersection of events whose probabilities are defined as $P(TTCT_i < t)$ where $i = 1, NCT$. It states that the complex task under consideration will be completed if, and only if, all the component tasks are completed. In the case that random variables $TTCT_i$, $i = 1, NCT$ represent independent events, the above expression becomes

$$\mathcal{M}(t) = P(TTCT_1 < t) \times P(TTCT_2 < t) \times \ldots \times P(TTCT_{NCT} < t)$$
$$= \prod_{i=1}^{NCT} P(TTCT_i < t)$$

According to Eq. (8.1), the expression on the right is the maintainability function of ith task. Consequently, the maintainability function of the complex task which consists of any number of tasks NCT, can be calculated according to the following equation

$$\mathcal{M}(t) = \prod_{i=1}^{NCT} M_i(t) \qquad (12.1)$$

where $M_i(t)$ is the maintainability function of ith task.

The right-hand side of the above expression represents the product of the probability functions of activities and is known as the product. It is necessary to stress that the above expression can be very complex. The reason for this is that the maintainability functions of the consisting tasks, $M_i(t)$ where $i = 1, NCT$, can be defined by any of the theoretical probability distributions and in the majority of cases their products cannot be readily expressed. For example, if two activities are performed simultaneously and their maintainability functions are expressed by, say, the normal or Weibull distribution, the maintainability function of the task as a whole cannot be defined. The situation gets worse as the number of activities increases. Therefore, Eq. (12.1) defines the maintainability function of the complex task at discrete points of time, as the product of the maintainability functions of consisting tasks. However, this does not define the probability distribution of a random variable TTR in an analytical form.

Analysing Eq. (12.1), it can be deduced that the probability of the task being complete by a specific instant of time will always be less than or equal to the lowest probability of all of the consisting tasks, thus

$$\mathcal{M}(t) < \{M_i(t), \quad i = 1, NCT\}_{min}$$

Table 12.1 clearly illustrates the point made by the above expression.

Example 12.1 A complex maintenance task consists of four tasks, A_1, A_2, A_3 and A_4, performed simultaneously. The maintainability functions for the tasks are defined: exponential for $A_1, E(10)$; Weibull for A_2, W(12, 3.2); normal for task A_3 N(8, 3.5); and Weibull for task A_4, W(20, 1.75). All parameters are in hours. Derive the expression for the maintainability function of the task, $\mathcal{M}(t)$, and calculate the probability of the task being complete after the first 10, 15 and 20 hours of restoration.

Table 12.1 Maintainability influenced by number of consisting tasks

NCT in the task	$M_i(t)$	$\mathcal{M}_s(t)$
2	0.9	0.81
3	0.9	0.729
4	0.9	0.656
5	0.9	0.59

SOLUTION According to the information given and in Table 8.1, the maintainability functions of the consisting tasks can be expressed as

$$M_{A_1}(t) = 1 - \exp\left(-\frac{t}{10}\right)$$

$$M_{A_2}(t) = 1 - \exp\left[-\left(\frac{t}{12}\right)^{3.2}\right]$$

$$M_{A_3}(t) = \int_0^t \frac{1}{3.5\sqrt{2\pi}} \exp\left[-\frac{1}{2}\left(\frac{t-8}{3.5}\right)^2\right] = \Phi\left(\frac{t-8}{3.5}\right)$$

$$M_{A_4}(t) = 1 - \exp\left[-\left(\frac{t}{20}\right)^{1.75}\right]$$

As all activities are performed simultaneously the maintainability function of the task is defined by Eq. (12.1), thus

$$M_t(t) = M_{A_1}(t) \times M_{A_2}(t) \times M_{A_3}(t) \times M_{A_4}(t)$$

$$= 1 - \exp\left(-\frac{t}{10}\right)$$

$$\times 1 - \exp\left[-\left(\frac{t}{12}\right)^{3.2}\right]$$

$$\times \int_t^\infty \frac{1}{3.5\sqrt{2\pi}} \exp\left[-\frac{1}{2}\left(\frac{t-8}{3.5}\right)^2\right]$$

$$\times 1 - \exp\left[-\left(\frac{t}{20}\right)^{1.75}\right]$$

Although the above expression is not easy to use, it fully defines the probability of the task being completed at any instant of time, $0 < TTR < \infty$. In providing solutions the random variable $TTCT_i$ will take specific values, as shown in Table 12.2. Figure 12.1 shows the maintainability function of the complex task, as well as the maintainability functions of consisting tasks.

Table 12.2 Maintainability values for Example 12.1

t	$M_{A_1}(t)$	$M_{A_2}(t)$	$M_{A_3}(t)$	$M_{A_4}(t)$	$\mathcal{M}(t)$
10	0.63212	0.427637	0.7161797	0.25718	0.049789
15	0.776869	0.87026	0.97725	0.45362	0.29971
20	0.86466	0.994069	0.99969	0.63212	0.543166

The maintainability function has been derived. The main objective now is to derive expressions for other maintainability measures for the complex task, based on the maintainability function, $\mathcal{M}(t)$.

12.2.1 Prediction of TTR_p time

This is the length of maintenance time by which the functionability of a given percentage of a population will be restored. It is the abscissa

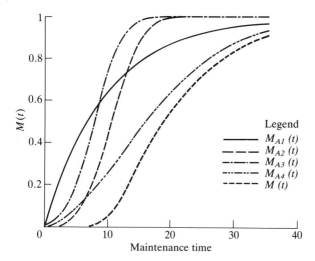

Figure 12.1 Maintainability functions for Example 12.1.

of the point whose ordinate presents a given percentage of restoration. Mathematically, TTR_p time can be represented as

$$TTR_p = t \longrightarrow \text{ for which } \mathcal{M}(t) = P(TTR \le t) = \int_0^t m(t)dt = p \quad (12.2)$$

where $\mathcal{M}(t)$ is the maintainability function of the complex task analysed.

Example 12.2 For the data given in Example 12.1, determine the maintenance time by which 85 per cent of complex tasks will be complete.

SOLUTION According to Eq. (12.2) determine the value of t for which $\mathcal{M}(t) = 0.85$, thus

$$TTR_{85} = t \longrightarrow \text{ for which } \mathcal{M}(t) = P(TTR \le t) = \int_0^t m(t)dt = 0.85$$

where the maintainability function $\mathcal{M}(t)$ has been derived in Example 12.1. Looking at this very complex expression it seems that the solution can be found more easily by using a graphical presentation. According to Fig. 12.2 the maintenance time by which 85 per cent of maintenance tasks will be complete is about 31 hours.

12.2.2 Prediction of expected restoration time

Making use of Eq. (8.3), the expression for the expected restoration time, $E(TTR)$, or mean time to restore, $MTTR$, can be derived as

$$E(TTR) = MTTR = \int_0^\infty t \times m(t)dt \quad (12.3)$$

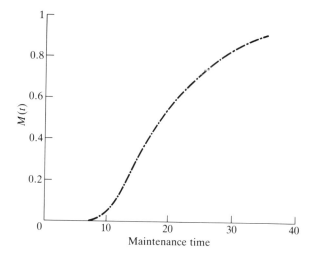

Figure 12.2 Maintainability function for Example 12.2.

Making use of Eq. (5.21), the above expression can be transformed into the following form

$$E(TTR) = MTTR = \int_0^\infty [1 - \mathcal{M}(t)]dt \qquad (12.4)$$

Generally speaking, the easiest way of calculating the numerical value of $MTTR$ is to apply numerical integration.

> **Example 12.3** Determine the mean time to restore functionability to the system specified in Example 12.1.
>
> SOLUTION The numerical value of $MTTR$ can be calculated by employing Eq. (12.4), where the expression for $\mathcal{M}(t)$ has been derived in Example 12.1. Clearly, integration of the expression represents an almost unsolvable problem. Hence, the required value will be determined by using numerical integration. In essence, the numerical value of $MTTR$ is equal to the area below the curve $1 - \mathcal{M}(t)$ as shown in Fig. 12.3. Hence, the mean time to failure, or expected value of random variable TTR, is around 21 hours.

12.2.3 Restoration success

The probability that the complex maintenance task under consideration, defined by the maintainability function $\mathcal{M}(t)$, has not been completed by the instant of time t_1 but will be completed by the instant of time t_2 can be determined by applying Eq. (8.5)

$$\begin{aligned} \mathcal{RS}_s(t_1, t_2) &= P(TTR < t_2 | TTR > t_1) \\ &= \frac{\mathcal{M}(t_2) - \mathcal{M}(t_1)}{1 - \mathcal{M}(t_1)} \end{aligned} \qquad (12.5)$$

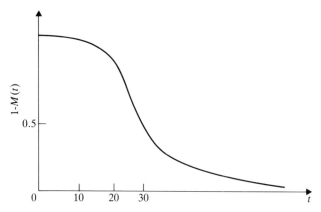

Figure 12.3 Complementary function to maintainability, Example 12.3.

Equation 12.5 represents the probability that completion of the complex maintenance task will occur within the next time interval.

Example 12.4 Determine the probability of completing the maintenance task within 15 and 20 hours given that the restoration process has not been completed by the end of the first 10 hours. The configuration of the task and its maintainability function are presented in Example 12.1.

SOLUTION The numerical values of the maintainability function are given in Table 12.1. According to Eq. (12.5), the mission maintainability of the task under consideration will be

$$\mathcal{M}\mathcal{S}_s(10, 15) = \frac{\mathcal{M}(15) - \mathcal{M}(10)}{1 - \mathcal{M}(10)} = \frac{0.29971 - 0.049789}{1 - 0.049789} = 0.249921/0.950211 = 0.263$$

Thus, there is a 26 per cent chance that the complex maintenance task will be completed within a 5-hour period of maintenance, beginning after the first 10 hours of restoration.

$$\mathcal{M}\mathcal{S}_s(10, 20) = \frac{\mathcal{M}(20) - \mathcal{M}(10)}{1 - \mathcal{M}(10)} = \frac{0.543166 - 0.049789}{1 - 0.049789} = 0.493377/0.950211 = 0.519$$

A complex task which has not been completed within the first 10 hours of restoration has a 51 per cent chance of being completed within the following 10 hours.

THIRTEEN

SUPPORTABILITY PREDICTION

The main objective of this chapter is to analyse a support task as a set of support activities and to derive expressions for the prediction of supportability measures based on corresponding measures for the consisting activities.

In Chapter 4 the concept of supportability was introduced and in Chapter 9 the measures which numerically express the ability of the item to be supported through a support task were described.

Using the notation in Chapter 9 the following variables will be introduced:

- $TTCSA_i$—random variable for the time to complete the ith support activity.
- TTS—random variable for the time to complete the support task as a whole.

The fundamental question to be posed here is how to predict supportability measures for the specific support task if the supportability measures of the consisting activities are known.

13.1 SUPPORT TASK

A complete support task is defined as:

A set of simultaneously performed support activities all of which must be completed if the task is to be completed.

The definition fully describes the relationship between consisting activities, clearly stating that:

(a) All activities start at the same instant of time and they are performed simultaneously.
(b) The failure to complete any of the consisting activities causes the failure of the task as a whole.

A typical example of a support task is the provision of support resources for the overhaul of the engine. This consists of support activities to provide spare parts, materials, technical data, personnel, tools, equipment and facilities as required. All of these activities have to be completed in order for the overhaul to be performed.

13.2 PREDICTION OF SUPPORTABILITY MEASURES FOR TASKS

Supportability measures of the task whose consisting activities are performed simultaneously can be derived from the supportability measures of its consisting activities.

13.2.1 Prediction of supportability functions

The supportability function for the support task, $\mathscr{S}(t)$, whose consisting activities are performed simultaneously, represents the probability that all support resources will be provided by a certain instant of time, t. This is defined as

$$\mathscr{S}(t) = P(TTS < t)$$
$$= P(TTCSA_1 < t \cap TTCSA_2 < t \cap TTCSA_3$$
$$< t \cap .... \cap TTCSA_{NCA} < t)$$

where NCA is the number of consisting activities involved. This expression mathematically describes the support task as an intersection of events whose probabilities are defined as $P(TTCSA_i < t)$, where $i = 1, NCA$. It clearly states that the task under consideration will be complete if, and only if, each of the consisting activities are complete.

Assuming that the random variables $TTCSA_i$, $i = 1, NCA$ represent independent events the above expression becomes

$$\mathscr{S}(t) = P(TTCSA_1 < t) \times P(TTCSA_2 < t) \times ... \times P(TTCSA_n < t)$$
$$= \prod_{i=1}^{NCA} P(TTCSA_i < t)$$

According to Eq. (9.1), the expression on the right-hand side is the supportability function for the i th activity. Consequently, the supportability task, which comprises any number of consisting activities NCA, can be calculated according to the following equation

$$\mathscr{S}(t) = \prod_{i=1}^{NCA} S_i(t) \tag{13.1}$$

where $S_i(t)$ is the supportability function for i th activity. The right-hand side of the above expression represents the product of supportability activities known as the product rule.

It is necessary to emphasize that the above expression can be very complex. This is because the supportability functions for the consisting activities, $S_i(t)$ where $i = 1, NCA$ can be defined through any of the theoretical probability distributions whereas, in the majority of cases, their products cannot be so expressed. For example, if two activities are performed simultaneously and their supportability functions are expressed with, say, the lognormal and Weibull distributions, the supportability function of the task as a whole cannot be defined by these distributions. The situation gets worse as the number of activities increases. Therefore, Eq. (13.1) defines the supportability function of the task at discrete points of time, as the product of the supportability functions for the consisting activities at that time. However, this does not define the probability distribution of the random variable TTS in analytical form.

Analysis of Eq. (13.1) shows that the probability of the task being completed by a specified instant of time will always be less than or equal to the lowest probability of the consisting activities, thus

$$\mathscr{S}(t) < \{S_i(t), \quad i = 1, NCA\}_{min}$$

Table 13.1 clearly illustrates the point made by the above expression.

Example 13.1 A support task consists of four activities, A_1, A_2, A_3 and A_4, performed simultaneously. The supportability functions for the activities are: exponential for $A_1, E(100)$; Weibull for $A_2, W(120, 3.2)$; normal for $A_3, N(80, 35)$; and Weibull for $A_4, W(200, 1.75)$. All parameters are in hours. Derive the expression for the supportability

Table 13.1 Effect on supportability by the number of consisting activities

NCA in the task	$S_i(t)$	$\mathscr{S}(t)$
2	0.99	0.9801
3	0.99	0.9703
4	0.99	0.9606
5	0.99	0.951

Table 13.2 Supportability values for Example 13.1

t	$S_{A_1}(t)$	$S_{A_2}(t)$	$S_{A_3}(t)$	$S_{A_4}(t)$	$\mathscr{S}(t)$
175	0.8262	0.9647	0.9966	0.5468	0.4344
200	0.8646	0.9941	0.9997	0.6321	0.5432
250	0.9179	0.9999	0.9999	0.7718	0.7084

function of the task, $\mathscr{S}(t)$, and calculate the probability of the task being completed after the first 175, 200 and 250 hours of restoration.

SOLUTION According to the information given and in Table 9.1, the supportability functions of the consisting activities can be expressed

$$S_{A_1}(t) = 1 - \exp\left(-\frac{t}{100}\right)$$

$$S_{A_2}(t) = 1 - \exp\left[-\left(\frac{t}{120}\right)^{3.2}\right]$$

$$S_{A_3}(t) = \int_0^t \frac{1}{35\sqrt{2\pi}} \exp\left[-\frac{1}{2}\left(\frac{t-80}{35}\right)^2\right]$$

$$= \Phi\left(\frac{t-80}{35}\right)$$

$$S_{A_4}(t) = 1 - \exp\left[-\left(\frac{t}{200}\right)^{1.75}\right]$$

As all activities are performed simultaneously, the supportability function for the task is defined by Eq. (13.1), thus

$$S_t(t) = S_{A_1}(t) \times S_{A_2}(t) \times S_{A_3}(t) \times S_{A_4}(t)$$

$$= 1 - \exp\left(-\frac{t}{100}\right)$$

$$\times 1 - \exp\left[-\left(\frac{t}{120}\right)^{3.2}\right]$$

$$\times \int_0^t \frac{1}{35\sqrt{2\pi}} \exp\left[-\frac{1}{2}\left(\frac{t-80}{35}\right)^2\right]$$

$$\times 1 - \exp\left[-\left(\frac{t}{200}\right)^{1.75}\right]$$

Clearly the above expression is not easy to use, but it fully defines the probability of the task being completed at any instant of time, $0 < TTS < \infty$. By way of solution the random variable $TTCSA_i$ will take specific values, as shown in Table 13.2. Figure 13.1 further illustrates this example.

13.2.2 Prediction of TTS_p time

This is the time by which the support resources will be provided by a given percentage of support tasks. It is the abscissa of the point whose ordinate

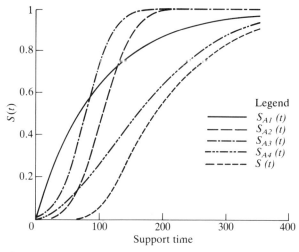

Figure 13.1 Supportability functions for Example 13.1.

presents a given percentage of completion. Mathematically, TTS_p time can be represented as

$$TTS_p = t \longrightarrow \text{ for which } \mathscr{S}(t) = P(TTR \le t) = \int_0^t s(t)dt = p \qquad (13.2)$$

where $\mathscr{S}(t)$ is the supportability function of the task analysed.

Example 13.2 For the data given in Example 13.1, determine the length of the support time by which 75 per cent of tasks will be completed.

SOLUTION According to Eq. (13.2) the required value for t is that one which satisfies $\mathscr{S}(t) = 0.75$, thus

$$TTS_{75} = t \longrightarrow \text{ for which } \mathscr{S}(t) = P(TTS \le t) = \int_0^t s(t)dt = 0.75$$

where the supportability function $\mathscr{S}(t)$ has been derived in the example above. This expression suggests that the solution could be found more easily by using a graphical representation of this function and, by reference to Fig. 13.2, the support time by which 75 per cent of support tasks will be completed is about 266 hours.

13.2.3 Prediction of expected support time

Making use of Eq. (9.3), the expression for the expected support time, $E(TTS)$, or mean time to support, $MTTS$, can be derived as

$$E(TTS) = MTTS = \int_0^\infty t \times s(t)dt \qquad (13.3)$$

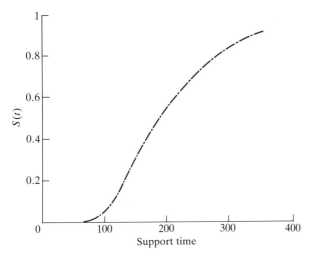

Figure 13.2 Supportability function for Example 13.2.

Using Eq. (5.21), the above expression can be transformed into the following form

$$E(TTS) = MTTS = \int_0^\infty [1 - \mathscr{S}(t)]dt \qquad (13.4)$$

Generally speaking, the easiest way of calculating the numerical value of $MTTS$ is to apply numerical integration.

Example 13.3 Determine the mean time to support the maintenance task of the system analysed in Example 13.1.

SOLUTION The numerical value of $MTTS$ can be calculated by making use of Eq. (13.4), where the expression for $\mathscr{S}(t)$ has been derived, in Example 13.1. Clearly, the integration of that expression represents an almost unsoluble problem. Hence, the required value will be determined by numerical integration. In essence, the numerical value of $MTTS$ is equal to the area below the curve $1 - \mathscr{S}(t)$, shown in Fig. 13.3. Hence, the mean time to support, or the expected value of the random variable TTS, is around 207 hours.

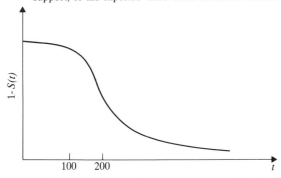

Figure 13.3 Complementary function to supportability, Example 13.3.

13.2.4 Support success

The probability that the support task under consideration, defined by the supportability function $\mathscr{S}_t(t)$, which has not been completed by the instant of time t_1 and will be completed by the instant of time t_2, can be determined by applying Eq. (9.5), thus

$$\mathscr{S}\mathscr{S}_s(t_1, t_2) = P(TTS < t_2 | TTS > t_1)$$
$$= \frac{\mathscr{S}(t_2) - \mathscr{S}(t_1)}{1 - \mathscr{S}(t_1)} \quad (13.5)$$

This expression represents the probability of completion of the support task within the specified time interval (given that the task under consideration has not been completed at the beginning of that interval).

Example 13.4 Determine the probability of completing the support task within 24 hours given that the support process has not been completed by the end of the first 210 hours. The specification of the task and its supportability function are derived in Example 13.1.

SOLUTION The numerical values for the supportability function are given in Table 13.1. According to Eq. (13.5), support success of the task will be

$$\mathscr{S}\mathscr{S}_s(210, 234) = \frac{\mathscr{S}(234) - \mathscr{S}(210)}{1 - \mathscr{S}(210)} = \frac{0.6612 - 0.58073}{1 - 0.58073} = 0.08047/0.41927 = 0.192$$

Thus, there is a 19 per cent chance that the support task will be completed within the next 24 hours.

FOURTEEN

PROBABILITY SYSTEM AND EMPIRICAL DATA

Analysing measures for reliability, R, maintainability, M, and supportability, S, derived in previous chapters, it is not difficult to deduce that these are all fully defined by the probability system, and in particular by the probability distribution of the appropriate random variable. Thus, measures derived for R, M, and S are based on one, or a combination of:

- Probability density function, $f(.)$
- Cumulative distribution function, $F(.)$
- Expected value, $E(.)$
- Variance, $V(.)$.

See Figs 7.1, 8.1 and 9.1. In order to reinforce this statement Table 14.1 gives the main measures through which reliability, maintainability and supportability characteristics, for an item or a system, can be expressed. At the same time all these characteristics are fully defined by the type of distribution function (see Tables 7.2, 7.3, 7.4, 8.1, 8.2, 9.1, and 9.2).

Table 14.1 Measures of reliability, maintainability and supportability

	Reliability	Maintainability	Supportability
Measure	$TSoFu$	$TSoFa$	$TSoFa$
Random variable	TTF	TTR	TTS
Probability density	$f(t)$	$m(t)$	$s(t)$
Cumulative probability	$F(t)$	$M(t)$	$S(t)$
Expected value	$E(TTF) = MTTF$	$E(TTR) = MTTR$	$E(TTS) = MTTS$
Percentage time	TTF_p	TTR_p	TTS_p

In Chapter 6 it has been shown that all the well-known distribution functions (exponential, normal, lognormal, and Weibull), are derived theoretically and are reproducible only if the experiment is repeated an infinite number of times. On the other hand, looking at examples from everyday engineering practice it is not difficult to conclude that in the majority of cases the family (and the particular member of the family) of distribution functions which could be used to describe the behaviour of the relevant random variable is unknown. This presents a serious problem for engineers because the empirical data related to the reliability of engineering items are usually limited. This is due to the fact that repetition of experiments is both expensive and time consuming (sometimes years and decades). The situation is very similar (and even worse) with empirical data that relate to measures of maintainability and supportability.

Thus, the problem which engineers face is the selection of the appropriate distribution function to describe the outcome of an experiment using the empirical data obtained from only a few repetitions. Expressed in the language of statistics the task is to make inferences about a population, based on the information contained in a sample.

The main concern of this chapter is therefore the presentation of the methodologies for establishing the relationship between the abstract probability system and probability distributions, created by mathematicians, and the empirical data related to reliability, maintainability and supportability measures, obtained in day-to-day engineering practice. In other words we are concerned with the selection of the family of theoretical probability distributions and associated parameters which define the random variables under consideration, (TTF, TTR, TTS), based on empirical data obtained from the repetition of real world processes.

This chapter describes some well-known methods for selection of the most relevant theoretical distribution functions for random variables. The overall process consists of a few independent but interconnected phases, as shown in Fig. 14.1. Each phase will be explained theoretically and illustrated with examples related to the most frequently used theoretical probability distributions, applicable to engineering practice.

14.1 PROBABILITY DISTRIBUTION OF EMPIRICAL DATA

According to the algorithm given in Fig. 14.1, the first step is the estimation of the probability distribution of empirical data obtained from the experiment performed. It is necessary to emphasize that the term experiment covers a large number of activities whose aim is to generate and collect raw data (laboratory test, field data, user reports, demonstration test and similar). Regarding the total number of results available there are two dis-

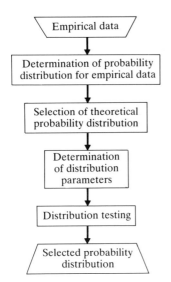

Figure 14.1 Algorithm for the selection of theoretical distribution functions for empirical data.

tinguishable procedures for the determination of the probability distribution of empirical data:

- Ranking method
- Relevant frequency method.

The first one is applicable in all cases where the total number of empirical data is less than 50, whereas the second method is recommendable in all other situations. Both methods are described below for a hypothetical random variable X.

14.1.1 Number of empirical data less than fifty

According to Bernoulli's theorem (Chapter 5, Eq. 5.9) the probability of occurrence of an event is defined by the ratio between its frequency and the total number of results, n, when this approaches infinity. The problem with the above statement starts in cases where the total number of results is insufficient for the data to be classified into groups. For example, if an experiment has been repeated four times the result with the lowest value, say x_1, has, from the point of view of cumulative distribution function, the probability of occurrence 0.25. On the other hand if the same experiment is repeated eight times, then the result with the lowest value x_1 represents the cumulative probability of 0.125 (1/8), and the second lowest value represents the cumulative probability of 0.25. It is clear that the difference between the proportion of data represented by the cumulative probabilities in these two cases directly depends on the total number of results. Similarly the cumulative distribution function for the result with the highest numerical value will be 1 (4/4). Obviously, this is not correct because there is always some possibility that a higher value could be obtained if the experiment were repeated more than four times.

Attempts have been made to solve this problem, and the ranking method seems to be the most suitable. The median rank, MR, which represents the median of all possible results which could have occurred in specific rank, is the most frequently used ranking method. In some cases the mean rank method can be used as well. Numerical values of median rank for all cases, where the total number of results is less than or equal to 50, are presented in Table T2. According to this table, for $n = 4$, the median rank of the first result is 0.1591 and not 0.25 as might have been expected, the second is 0.38636, the third 0.61364 and the fourth 0.84091.

Apart from using the available tables, the approximate numerical values of the median rank can be determined by using the following expression

$$MR = \frac{(i - 0.3)}{(n + 0.4)} \tag{14.1}$$

where i is the order number of each individual result after they have been arranged in ascending order, and n is the total number of results available. The above approximation is accurate to 1 per cent for $n > 5$ and to 0.1 per cent for $n > 50$.

Other than the median rank, the most commonly used are 5 and 95 per cent ranks which correspond to probabilities of 0.05 and 0.95. Table T2 shows that in the case of four results, the 5 per cent rank has a value of 0.01274 which means that the probability of 0.05 exists that the first result will have cumulative probability greater than 0.01274. Thus in only 5 per cent of cases will the first result represent more than 1.27 per cent of the total results. For the same example, rank 0.95 is 0.52713 which means that in 95 per cent of cases, the first results will present less than 52.7 per cent of the total number of results.

As the median rank represents the cumulative proportion of the data corresponding to the rank under consideration, the cumulative distribution function of empirical data, $F'(.)$ can be determined in the same way, thus for the random variable X

$$F'(x_i) = MR(x_i) = \frac{(i - 0.3)}{(n + 0.4)} \tag{14.2}$$

By making use of the above equation the expression for the probability density function of empirical data, $f'(.)$, can be derived as

$$f'(x_i) = \frac{1}{(n + 0.4) \times (x_{i+1} - x_i)} \tag{14.3}$$

It is necessary to repeat that, in order to apply the above expressions to the empirical data, they must be arranged in ascending order, i.e. $i = 1$ corresponds to the result with the lowest numerical value.

In the same way as the probability distribution of the random variable could be represented by the summary measures, the probability distribution of the empirical data could be described by the measures of central tendency and the measures of dispersion. Thus, the mean value of empirical data, M' can be determined by the expression

$$M' = \sum_{i=1}^{n} \frac{x_i}{n} \tag{14.4}$$

The median, $M'd$, in the case of an odd number of results is equal to the numerical value of the middle result, whereas in the case of even numbers of results, the median is equal to the arithmetical mean of the middle two results.

The standard deviation, SD', can be calculated

$$SD' = \sqrt{\frac{\sum\limits_{i=1}^{n} (x_i - M')^2}{n-1}} \tag{14.5}$$

In order to clarify the above statements and make the calculation easier, the form named R1 was designed. It contains all the steps for the determination of probability characteristics based on the above equations. The form shown in Fig. 14.2 is applicable to all random variables.

Example 14.1 The time to failure of nine items tested are presented below. Calculate and plot the probability characteristics of the empirical data available.

result number	1	2	3	4	5	6	7	8	9
measured value	84	34	37	99	51	69	26	48	103

SOLUTION The first step towards the solution is to rearrange the available data into ascending order:

order number	1	2	3	4	5	6	7	8	9
measured value	26	34	37	48	51	69	84	99	103

In order to calculate the mean and standard deviation of empirical data Eqs (14.4) and (14.5) should be used.

Regarding the determination of distribution functions for the empirical data, the following two ways are possible:

(a) Use the data given in Table T2, in the following way: find the part which is applicable to sample size $n = 9$ and directly read values for $MR(i) = F'(x_i)$ for $i = 1, 9$. For example, $F'(x_1) = 0.06697$ (6.7 per cent). Then the numerical values of $f'(x_i)$ can be calculated by applying Eq. (14.3).
(b) Use form R1, as shown in Fig. 14.3.

Graphical interpretations of the obtained functions are presented in Fig. 14.4.

Example 14.1 illustrates a commonly accepted procedure for the calculation of probability distributions for empirical data when the total number of results is less than 50 ($n < 50$). It should be emphasized that extreme care must be taken since a single inaccurate result can distort the picture, by changing the shape of the functions.

Total Number of Results, n = ...		Order Number of results, i = 1,n		B = n + 0.4 = ... + 0.4 = ...	
i	x_i	$(x_i - M')^2$	$F'(x_i) = (i - 0.3)/B$	$C_i = x_{i+1} - x_i$	$f'(x_1) = 1/C_i B$
$\sum_{i+1}^{n}$	S1	S2			

$$M' = S1/n \ \text{............} = \text{............}$$

$$SD' = \sqrt{S2/(n-1)} = \sqrt{(\text{............})/(\text{............} -1)} = \text{............}$$

Figure 14.2 Form R1.

Total Number of Results, n = ...9....			Order Number of results, i = 1,n	B = n + 0.4 = ...9.... + 0.4 = ...9.4....	
i	x_i	$(x_i - M')^2$	$F'(x_i) = (i - 0.3)/B$	$C_i = x_{i+1} - x_i$	$t'(x_i) = 1/C_i \cdot B$
1	26	1240.45	0.074	8	0.013298
2	34	740.9	0.181	3	0.035461
3	37	586.6	0.287	11	0.009671
4	48	174.9	0.394	3	0.035461
5	51	104.4	0.500	18	0.005910
6	69	60.52	0.606	15	0.007092
7	84	518.9	0.713	15	0.007092
8	99	1427.3	0.819	4	0.026596
9	103	1745.5	0.926		
$\sum_{i+1}^{n}$	551	6599.5			
	S1	S2			

$$M' = S1/n \quad ....551..../....9.... = ...61.22....$$

$$SD' = \sqrt{S2/(n-1)} = \sqrt{(....6599.5....)/(....9-1....)} = ...28.72....$$

Figure 14.3 Form R1 with data related to Example 14.1.

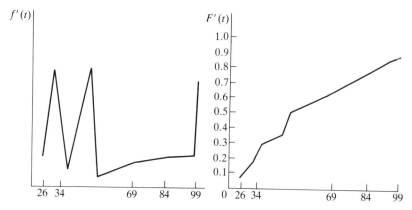

Figure 14.4 *CDF'* and *PDF'* for empirical data.

14.1.2 Number of empirical data greater than fifty

In contrast to the method above, where results had to be ranked, in this case it is possible to use the frequency approach and Eq. (5.9). In order to calculate the probability characteristics of the empirical data the results are classified into several intervals the number of which, *NI*, depends on the total number of results *n*. The following equation can be used as guidance for determining the suitable number of intervals, thus

$$NI = 1 + 3.3 \times log_{10}(n) \qquad (14.6)$$

The length of each interval, *LI*, is defined by

$$LI = \frac{(x_n - x_1)}{NI} \qquad (14.7)$$

The limits of the intervals can be calculated in the following way

$$x_{min_i} = x_i + (i - 1) \times LI$$
$$x_{max_i} = x_{min_i} + LI$$
$$x_{m_i} = \frac{x_{max_i} - x_{min_i}}{2}$$

where *i* is order number of interval, $i = 1, NI$; x_{min_i} is the minimum value qualifying for *i*th interval; x_{max_i} is the maximum qualifying value for *i*th interval and x_{m_i} is the mean value of *i*th interval.

Taking into account Eq. (14.1), the estimated values for the cumulative distribution function of the empirical data considered are defined by the expression

$$F'(x_{max_i}) = \frac{N(x_{max_i}) - 0.3}{n + 0.4} \qquad (14.8)$$

where $N(x_{max_i})$ is the total number of results whose numerical value is less than or equal to x_{max_i}.

The probability density function of empirical data is defined

$$f'(x_{m_i}) = \frac{f_i}{n \times LI} \tag{14.9}$$

where f_i is the frequency of ith interval, i.e. it is the number of results whose numerical values are greater than x_{min_i} and less than x_{max_i}.

The mean of empirical data can be obtained as

$$M' = \frac{\sum\limits_{i=1}^{NI} x_{m_i} \times f_i}{n} \tag{14.10}$$

The standard deviation of available data can be determined

$$SD' = \sqrt{\frac{\sum\limits_{i=1}^{NI} (x_{m_i} - M')^2 \times f_i}{n}} \tag{14.11}$$

and the same results can be obtained by using the transformed expression

$$SD' = \sqrt{\sum\limits_{i=1}^{NI} \frac{x_{m_i}^2}{n} \times f_i - M'^2} \tag{14.12}$$

In order to facilitate calculation of these numerical values in everyday practice, the form R2 was designed, which contains all the steps for the determination of the probability distribution for empirical data, Fig. 14.5.

Example 14.2 Results of 55 measured values from an experiment are given below

3	56	9	24	56	66	67	87	89	99	4
26	76	79	89	45	65	78	88	89	90	92
99	2	3	37	39	77	77	93	21	24	29
32	44	46	5	46	79	99	47	77	79	89
31	78	34	67	86	91	75	33	55	22	44

Calculate the mean and standard deviation of this empirical data which represents a random variable and plot the functions $F'(t)$ and $f'(t)$.

SOLUTION The form R2 was used to facilitate the calculation of the probability characteristics of the empirical data. The results obtained, after they have been classified into six groups, are given in Fig. 14.6. The graphical interpretation of the functions, $f(.)$ and $F(.)$ are shown as histograms in Fig. 14.7.

14.1.3 Algorithm for calculation

In order to provide a full picture of the method presented for the estimation of probability distribution of available empirical data, the algorithm shown in Fig. 14.8 was designed. It shows that the output characteristics of this

Total Number of Results, n = ... Number of Intervals, NI = 1+3.3 log (n) = ... Length of Interval, LI = $(x_n - x'_1)$/NI = ...

i	$x_{min_i} =$ $x_1 + (i-1)$ LI	$x_{max_i} =$ $x_{min_i} +$ LI	$x_{m_i} = (x_{min_i} +$ $x_{max_i})/2$	$x_{m_i}^2$	$f_i =$ $N(LI_i)$	$x_{m_i} f_i$	$x_{m_i}^2 f_i$	$b_i =$ $N(x_{max_i})$	$F'(x_{max_i}) = b_i/n$	$f'(x_{m_i}) = f_i/(n\,LI)$
$\sum_{i=1}^{NI}$						S1	S2			

$M' = S1/n = \ldots\ldots/\ldots\ldots = \ldots\ldots$

$SD' = \sqrt{S2/n - M'^2} = \sqrt{\ldots\ldots/\ldots\ldots - \ldots\ldots} = \ldots\ldots$

Bei16a

Figure 14.5 Form R2.

Total Number of Results, n = 55   Number of Intervals, NI = 1+3.3 log (n) = 1 + 33 log(55) = 6   Length of Interval, LI = $(x_n - x_1)$/NI = 16.17

i	$x_{min_i} =$ $x_1 + (i-1)$ LI	$x_{max_i} =$ x_{min_i} + LI	$x_{m_i} = (x_{min_i} +$ $x_{max_i})$/2	$x_{m_i}^2$	$f_i =$ $N(LI_i)$	$x_{m_i} f_i$	$x_{m_i}^2 f_i$	$b_i =$ $N(x_{max_i})$	$F'(x_{max_i}) = b_i/n$	$f'(x_{m_i}) = f_i/(n\,LI)$
1	2	18.17	10.08	101.61	6	60.48	609.64	6	0.10909	0.006746
2	18.17	34.33	26.25	689.06	10	262.5	6890.6	16	0.290909	0.011244
3	34.33	50.50	42.42	1799.45	8	339.3	14395.6	24	0.43636	0.00899
4	50.50	66.67	58.58	3431.6	5	292.9	17158.1	29	0.527272	0.005622
5	66.67	82.83	74.75	5587.5	12	897	67050.7	41	0.745454	0.013493
6	82.83	99.00	90.92	8266.4	14	1272.9	115730.2	55	1.0000	0.0157418
$\sum_{i=1}^{NI}$						3125.06	221834.8			

$$S1 = 3125.06 \qquad S2 = 221834.8$$

$$M' = S1/n = \frac{3125.06}{55} = 56.82$$

$$SD' = \sqrt{S2/n - M'^2} = \sqrt{\frac{221834.8}{55} - 56.82^2} = 28.37$$

Figure 14.6 Form R2 with data related to Example 14.2.

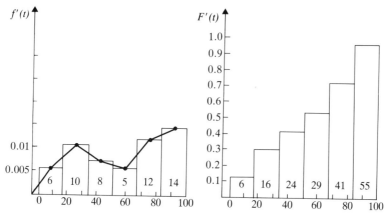

Figure 14.7 PDF' and CDF' of empirical data, Example 14.2.

phase are numerical values of the probability distribution for empirical data, presented in analytical and graphical form. Results obtained using this method must be treated with caution, because they represent only the available empirical data.

14.1.4 Suspended data

In practice a situation could arise where an experiment or process is suspended before its completion due to some unexpected cause. In these cases, especially when a limited number of results are available, the result achieved up to the moment of suspension must be taken into account, because we cannot afford to neglect this information. These cases are known as experiments with suspension and, for them, a special method for calculation of probability characteristics has been developed. Thus, in this case two types of data exist: final results denoted as F, and suspended, S. In this particular case the median rank method is also used but this time without the help of Table T2.

The procedure for dealing with suspended data is based on the following steps:

1. All empirical data should be organized in ascending order, marking suspended data.
2. For every final result after the suspended one, a real rank order should be determined by using the expression

$$I = \frac{(n+1) - (PON)}{1 + (NOR \text{ beyond present suspended results})} \qquad (14.13)$$

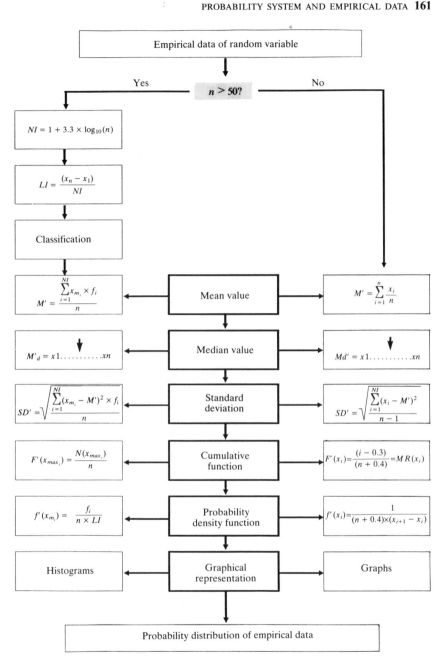

Figure 14.8 Algorithm for the calculation of probability distributions of empirical data.

where I represents increment order number, PON stands for the previous order number, and NOR represents the number of results.

The increment should be added to the previous rank order number until another suspended result is reached when a new increment should be calculated and the same procedure repeated. In general, the next final result will have a rank order number equal to the sum of preceding number increased and the new increment. Based on calculated new rank order for each final result, the probability characteristics of the empirical data can be determined by applying the method described above.

Example 14.3 Ten prototype motorbike struts and coil springs are tested on the proving ground section known as Belgian pave. The struts failed after the following distances (in km): 21 175, 28 033, 14 799, 30 102, 32 913, 23 871, 35 218, 25 338, 18 427, and 38 203. Note from the test recorded that the motorbike whose time to failure was 30 102 km was involved in a road accident so that testing was discontinued before the struts or coil springs had failed. The task is to determine the estimated values of the cumulative distribution function for the empirical data obtained by the test performed.

SOLUTION All the obtainable data are presented in Table 14.2.
The increment for result number 8 was calculated using Eq. (14.13), thus

$$I = [(10 + 1) - 6]/[1 + 3] = 1.25$$

This enabled us to allocate the order numbers shown in the table. After the sixth order number the next number is calculated as $6 + I = 6 + 1.25 = 7.25$, and next $7.25 + I = 7.25 + 1.25 = 8.50$, and so on.

In order to highlight the possible difficulties related to the practical application of the method described, the solutions for the three variations of the above example are shown in Tables 14.3, 14.4 and 14.5. Tables 14.6 and 14.7 are designed as a help to all practising engineers who are phasing the manipulation of empirical data whose total number is greater than 50, and the relative frequency method has to be applied.

Table 14.2 Empirical data for Example 14.3

Result number	Km	Type	Increment	Order number	Median rank
1	14 799	F	1	1	0.06731
2	18 427	F	1	2	0.16346
3	21 175	F	1	3	0.25962
4	23 871	F	1	4	0.35577
5	25 338	F	1	5	0.45192
6	28 033	F	1	6	0.54808
7	32 102	S	0	–	–
8	32 913	F	1.25	7.25	0.66827
9	35 218	F	1.25	8.50	0.78846
10	38 203	F	1.25	9.75	0.90865

Table 14.3 Example with three suspended data

Item number	Failure or suspension	Increment	Rank order
1	F	1.00	1
2	F	1.00	2
3	S	$\frac{10+1-2}{1+7}$	-
4	F	1.125	3.125
5	F	1.125	4.25
6	S	$\frac{10+1-4.25}{1+4}$	-
7	F	1.35	5.60
8	S	$\frac{10+1-5.60}{1+2}$	-
9	F	1.80	7.4
10	F	1.80	9.2

Table 14.4 Example with five suspended data

Item number	Failure or suspension	Increment	Rank order
1		1.00	1
2	F	1.00	2
3	S	$\frac{10+1-2}{1+7}$	-
4	S	$\frac{10+1-2}{1+6}$	-
5	S	$\frac{10+1-2}{1+5}$	-
6	F	1.50	3.5
7	F	1.50	5.0
8	S	$\frac{10+1-5.0}{1+2}$	-
9	S	$\frac{10+1-5.0}{1+1}$	-
10	F	3.0	8.0

14.2 SELECTION OF THEORETICAL DISTRIBUTION

In this phase, selection of the theoretical distribution function which represents the empirical data as closely as possible has to be performed. Using the language of statistics this process is known as hypothesis making.

It is very difficult to select one particular family of theoretical probability functions with which to associate the probability distribution of empirical data, especially in cases with a small number of results. The main indicator

Table 14.5 Example with seven suspended data

Item number	Failure or suspension	Increment	Rank order
1	S	$\frac{10+1-0}{1+9}$	-
2	S	$\frac{10+1-0}{1+8}$	-
3	S	$\frac{10+1-0}{1+7}$	-
4	F	1.375	1.375
5	S	$\frac{10+1-1.375}{1+5}$	-
6	S	$\frac{10+1-1.375}{1+4}$	-
7	S	$\frac{10+1-1.375}{1+3}$	-
8	S	$\frac{10+1-1.375}{1+2}$	-
9	F	3.208	4.583
10	F	3.208	7.792

for the selection of the theoretical probability distribution should come from information output from the previous step. That is, one or more of:

- Mean, M'
- Median, $M'd$
- Standard deviation, SD'
- Probability density function, $f'(x_i)$
- Cumulative distribution function, $F'(x_i)$

The following recommendations are provided to assist the decision-making process regarding the selection of the most suitable theoretical distribution for empirical data:

- If the estimated values of the standard deviation and the mean are relatively close, it is a good indicator that the random variable obeys exponential distribution.
- As the normal distribution is symmetrical, i.e. mean, median and mode have the same numerical value, in cases where the estimated values of $M'(.)$ and $Md'(.)$ are reasonably close, the recommended distribution is the normal.
- If the first two conditions are not met, and the plotted probability density function of the empirical data is not symmetrical it is most likely that we are dealing with a lognormal or Weibull distribution.
- If none of the above conditions are satisfied it is most likely that the empirical data does not belong to the same experiment or process, i.e.

Table 14.6 Example with large number of suspended data

Interval number	Number of failures	Number of suspensions	Increment	Rank order number	Median rank
1	3	7	$\frac{53+1-0}{1+46} = 1.15$	$1.15 \times 3 = 3.45$	0.0613
2	5	1	$\frac{53+1-3.45}{1+42} = 1.18$	$3.45 + 5 \times 1.18 = 9.33$	0.1691
3	17	8	$\frac{53+1-9.33}{1+29} = 1.49$	$9.33 + 17 \times 1.49 = 34.64$	0.6431
4	9	3	$\frac{53+1-34.64}{1+9} = 1.94$	$34.64 + 9 \times 1.94 = 52.06$	0.9693

Table 14.7 Example with large number of suspended data

Interval number	Number of failures	Number of suspensions	Increment	Rank order number	Median rank
1	0	5	$\frac{56+1-0}{1+51} = 1.10$	$1.10 \times 0 = 0.00$	0.0000
2	19	3	$\frac{56+1-0}{1+48} = 1.16$	$1.1 + 19 \times 1.16 = 22.10$	0.3865
3	14	0	$- = 1.16$	$22.10 + 14 \times 1.16 = 38.34$	0.6743
4	7	3	$\frac{56+1-38.34}{1+12} = 1.44$	$38.34 + 7 \times 1.44 = 48.39$	0.8527
5	0	2	$\frac{56+1-48.39}{1+3} = 2.15$	$48.30 + 0 \times 2.15 = 48.39$	0.8527
6	2	1	$\frac{56+1-48.39}{1+2} = 2.87$	$48.39 + 2 \times 2.87 = 54.13$	0.95444

that we are dealing with a mixed distribution. In this case the empirical data should be carefully re-examined and if there is evidence of a mixed distribution, it is necessary to separate mixed results, and then repeat the above procedure for each set of data.

When a hypothesis is made about the family of the theoretical distribution function it is necessary to find the numerical values of the scale parameter A, the shape parameter B and the source parameter C, which together fully define the member of the family. There are two possible ways of determining the parameters of theoretical distribution function for empirical data, graphical and analytical.

The analytical method requires more calculation time, but gives more accurate results, whereas the graphical method is simpler but leaves room for inaccuracy. In this chapter both methods will be described.

14.3 GRAPHICAL DETERMINATION OF DISTRIBUTION FUNCTION PARAMETERS

The basic tool for the graphical method for determination of the most suitable probability distribution is a special type of graph paper, known as *probability paper*. Probability paper for a hypothetical probability distribution is presented in Fig. 14.9. For each theoretical distribution function there is a unique probability paper whose coordinates are designed so that the probability distribution is given as a straight line. Thus, a cumulative distribution function, $F(x)$, defined by parameters A, B, C plotted on the appropriate probability paper, will be a straight line, with positive slope. Using this property, it is possible to plot empirical data on the probability paper, as a test of data distribution, and at the same time obtain a graphical assessment of the parameters involved by drawing a straight line to the empirical points.

All types of probability paper are used according to the following procedure:

1. *Plot the empirical data*—every point plotted on the probability paper is defined by the pair $x_i, F'(x_i)$, where $i = 1, n$, as shown in Fig. 14.10 on page 168.
2. *Visual check*—if the points plotted fall in a straight line, it indicates that the empirical data under consideration could be represented by a selected distribution. It should be said that it is unreasonable to expect a perfectly straight line, and that according to some authors the first and last 10 per cent of results can be ignored.

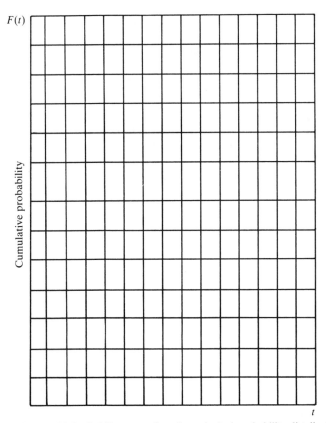

Figure 14.9 Probability paper for a hypothetical probability distribution.

3. *Fit the best line*—a straight line should be fitted through the plotted points. The chosen line should be the best representative of all points (see Fig. 14.11 page 169). This is the weakest part of the graphical method because the location of the line is subject to the discretion of the person involved.

This method for the selection of a probability function is simple and quick, but carries with it a possibility of error, due to the inaccurate fit of the straight line through the points. In some cases it is possible to fit several best lines as shown in Fig. 14.11. In short, subjective decision-making leaves room for error.

The procedure for determination of numerical values for the distribution parameters, A, B and C are different for each family of distribution. Hence, in the following sections, separate procedures for the exponential, normal, lognormal and Weibull distributions are given.

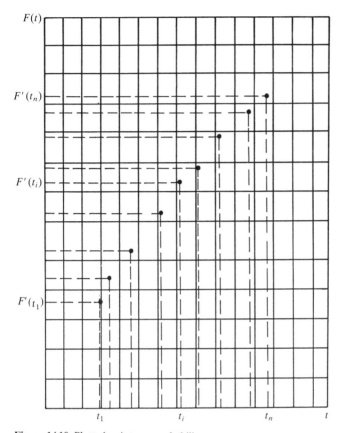

Figure 14.10 Plotted points on probability paper.

14.3.1 Exponential distribution

Probability paper for the exponential distribution is presented in Fig. 14.12 on page 170.

In order to determine the numerical value of the scale parameter, A, as the single parameter which defines the distribution, the following procedure is applied. From the point on the ordinate $F(x) = 0.632$ draw a horizontal line to the *best fit line* and then a vertical line down to the horizontal axis, where the numerical value for A can be read. This is shown in Fig. 14.13 (page 171) and is justified because the cumulative distribution function for the exponential distribution has a value of 0.632 for $x = A$, thus

$$F(A) = 1 - \exp[-(A/A)] = 1 - \exp(-1) = 0.632$$

The particular member of the exponential distribution family is fully defined by fixing the value of A.

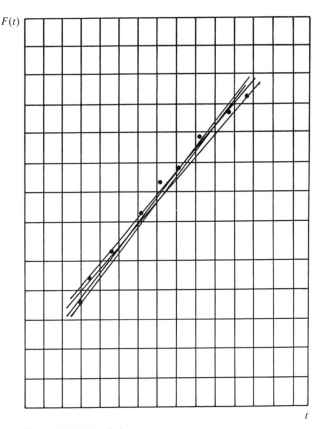

F(*t*)

t

Figure 14.11 Line fitting.

Example 14.4 The results tabulated represent the support time, in hours, needed for the provision of a specified spare part for 14 aeroplanes of a particular make and type.

i :	1	2	3	4	5	6	7	8	9	10	11	12	13	14
t_i :	102	209	14	57	54	32	67	134	152	27	230	66	61	34

Check the hypothesis of an exponential distribution, determine its parameter and plot the functions $f(t)$ and $F(t)$.

SOLUTION The first step towards solving this problem is to classify all results into ascending order (from minimum to maximum value). Then numerical values of $F'(t_i)$ should be determined. As the total number of results is less than 50, the median rank method must be used. The necessary values could be obtained either from Table T2, for $n = 14$, or calculated by applying Eq. (14.2). Values given in Table 14.8 are extracted from Table T2.

The points with coordinates $t_i, F'(t_i)$, $i = 1, 14$, plotted on probability paper for the exponential distribution are shown in Fig. 14.14 (see page 172). As the plotted points fall in a straight line there is no reason for the hypothesis to be rejected. After the best line was fitted through the points a numerical value of $A = 96$ hours is determined. Diagrams for $s(t)$ and $S(t)$, shown in Fig. 14.15 on page 172, are plotted by applying Eqs (6.9) and (6.10).

In conclusion, it can be said that:

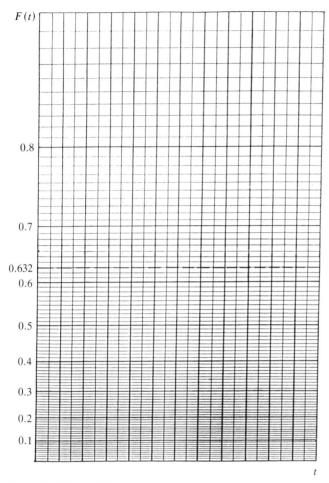

Figure 14.12 Probability paper for the exponential distribution.

(a) The hypothesis about the exponential distribution has been proved correct.

(b) The support time needed for the provision of specific spare parts for the particular type of aircraft could be represented by the exponential distribution with $A = 96$.

(c) It is possible to predict the support time needed for the provision of specified resources. For example, if we are interested to know what the probability is that a system will be supported within 59 hours the following expression could be used

$$P(T > 59) = 1 - P(T < 59) = 1 - \exp(-(59/96)) = 0.54$$

This means that there is a probability of 0.54 (54 per cent) that the required spares will be provided within 59 hours. This should not be accepted dogmatically, i.e. it should not be expected that exactly 46 systems out of 100 will be supported within 59 operating hours.

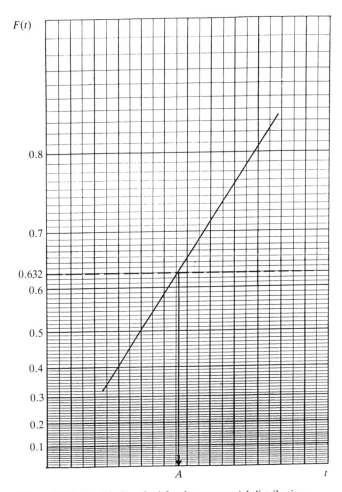

Figure 14.13 Graphical method for the exponential distribution.

Table 14.8 Values for Example 14.4

i	t_i	$F'(t_i)$	i	t_i	$F'(t_i)$
1	19.000	0.04861	8	66.000	0.53472
2	27.000	0.11806	9	67.000	0.60417
3	32.000	0.18750	10	102.000	0.67361
4	39.000	0.25694	11	139.000	0.74306
5	59.000	0.32639	12	152.000	0.81250
6	57.000	0.39583	13	209.000	0.88194
7	61.000	0.46528	14	230.000	0.95139

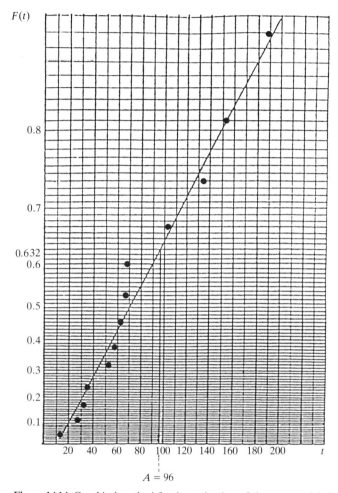

Figure 14.14 Graphical method for determination of the exponential distribution parameters.

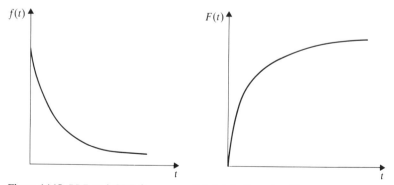

Figure 14.15 *PDF* and *CDF* for exponential distribution, $A = 96$.

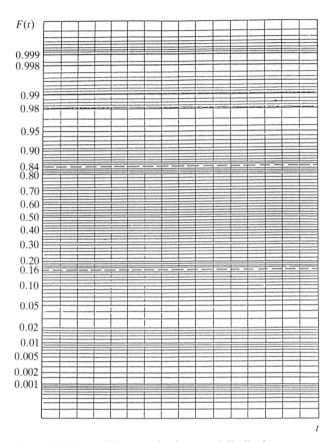

Figure 14.16 Probability paper for the normal distribution.

14.3.2 Normal distribution

Probability paper for the normal distribution is shown in Fig. 14.16. In the case of the normal distribution, the probability that the random variable will have a value less than or equal to scale parameter A is 0.5. Thus, $P(X \le A) = F(A) = 0.5$. The numerical value of the scale parameter for the empirical data under consideration, is the value of the point of intersection of the adopted best fit line and the horizontal line which corresponds to a cumulative probability of 0.5, denoted as $x_{0.5}$. The above statement can be mathematically expressed

$$A = x_{0.5} \quad \text{for which} \quad F(A) = 0.5 \tag{14.14}$$

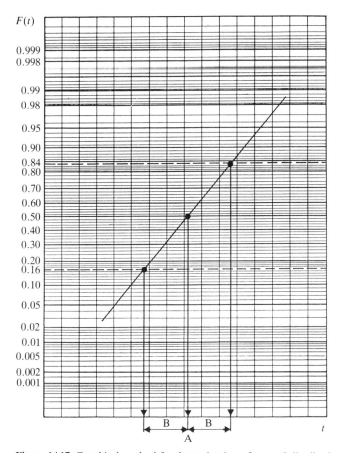

Figure 14.17 Graphical method for determination of normal distribution parameters.

The numerical value of the shape parameter, B, is equal to the difference between the values of abscissa whose ordinates have values 0.5 and 0.16 or 0.84 and 0.5. Thus

$$B = x_{0.50} - x_{0.16} = x_{0.84} - x_{0.50} \tag{14.15}$$

The procedure for determination of numerical values for parameters A and B, using the graphical method, is presented in Fig. 14.17.

Example 14.5 In a laboratory test of 19 washing machines, the operating time to failure of drive belts was recorded. The data obtained (operating hours) are presented below.

129 174 118 209 185 98 143 124 164 174 169 110 134 195 140 151 147 156 161

Assuming that the empirical data can be represented by the normal distribution, determine the parameters which define it and calculate the probability that failure will not occur in the first 125 hours.

Table 14.9 Values for Example 14.5

i	t_i	$F'(t_i)$	i	t_i	$F'(t_i)$
1	98.000	0.03608	11	156.000	0.55155
2	110.000	0.08763	12	161.000	0.60309
3	118.000	0.13918	13	169.000	0.65464
4	129.000	0.19072	14	169.000	0.70619
5	129.000	0.24227	15	179.000	0.75773
6	139.000	0.29381	16	179.000	0.80928
7	139.000	0.34536	17	186.000	0.86082
8	143.000	0.39691	18	195.000	0.91237
9	147.000	0.44845	19	209.000	0.96392
10	151.000	0.50000			

SOLUTION In order to solve the first part of this task the numerical values of the cumulative distribution function, $F'(t_i)$, for $i = 1, 19$ must be calculated. As the total number of results is less than 50, Eq. (14.2) should be used. The data calculated are presented in Table 14.9.

This data, plotted on probability paper for the normal distribution, is shown in Fig. 14.18 on page 176. As the data form a straight line a normal distribution can be assumed. Based on the position of the best line of fit parameters A and B are found to be $A = 152$, $B = 32$ hours. The cumulative distribution function can now be plotted by making use of Eq. (6.14), as shown in Fig. 14.19 on page 176.

The probability that the failure of drive belts will not occur in the first 125 hours of operation can be found by using the expression

$$P(T > 125) = 1 - P(T < 125) = 1 - F(125) = 1 - \Phi(\frac{125 - 152}{32}) = 0.8$$

The same result can be obtained using the function plotted in Fig. 14.19.

14.3.3 Lognormal distribution

Fig. 14.20 shows probability paper for the lognormal distribution. The similarity with probability paper for normal distribution is obvious. According to the description of the lognormal distribution given in Chapter 6 it is reasonable to expect that the process of determining numerical values for its parameters is similar to the process already explained for the normal distribution.

The points of interest are the numerical values of the X-axis which correspond to the numerical values of the intersection of the best line and the cumulative probabilities of 0.16, 0.5 and 0.84. Once these values are determined the distribution parameters can be obtained according to the following expressions

$$A = \ln(x_{0.50}) \tag{14.16}$$

$$B = \ln[1/2(x_{0.50}/x_{0.16} + x_{0.84}/x_{0.50})] \tag{14.17}$$

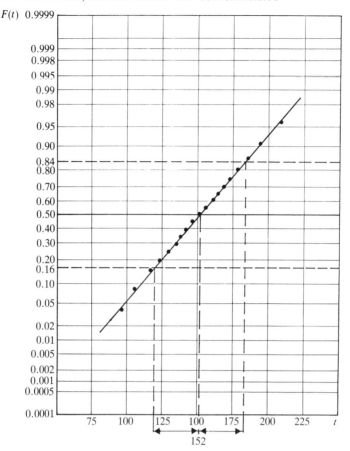

Figure 14.18 Graphical method for determination of the normal distribution parameters.

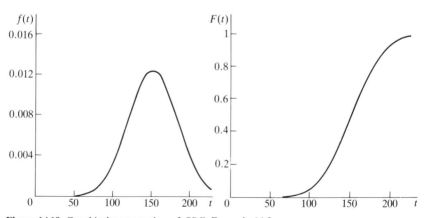

Figure 14.19 Graphical presentation of *CDF*, Example 14.5.

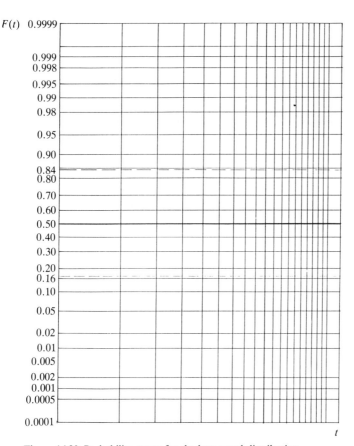

Figure 14.20 Probability paper for the lognormal distribution.

where $x_{0.50}$ is the value of x for which $F(x) = 0.5$, etc. The whole procedure is illustrated in Fig. 14.21.

Example 14.6 Five newly-designed engines were driven the equivalent of 30 000 miles. The following table represents the odometer readings at which the first failure of particular components occurred.

i :	1	2	3	4	5
t_i :	2600	3800	5100	7700	10 200

Can the above empirical data be represented by the lognormal distribution? If so, determine the expected value of operating time to the first component failure using the graphical method.

SOLUTION In order to check the above hypothesis, it is necessary to plot the points with coordinates $t_i, F'(t_i)$, $i = 1, 5$ on probability paper for the lognormal distribution. According to the distribution of the points shown in Fig. 14.22 there is no reason for rejecting the lognormal distribution. In order to calculate the mean value it is necessary

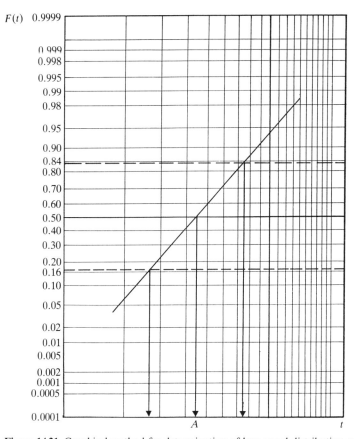

Figure 14.21 Graphical method for determination of lognormal distribution parameters.

to determine numerical values for parameters A and B. From Fig. 14.22, the numerical values obtained are

$$A = \ln(5150) = 8.546$$
$$B = \ln(1/2(5150/2970 + 9125/5150) = 0.561$$

Thus, the expected value of the operating time to failure of the component considered is

$$E(TTF) = \exp(A + 0.5B^2) = \exp(8.546 + 0.157) = 6025 \text{ miles}$$

14.3.4 Weibull distribution

In spite of the fact that several different probability papers for the Weibull distribution exist, they are all based on the same principle and all give the same values for parameters A, B and C. Only one type of paper is shown in

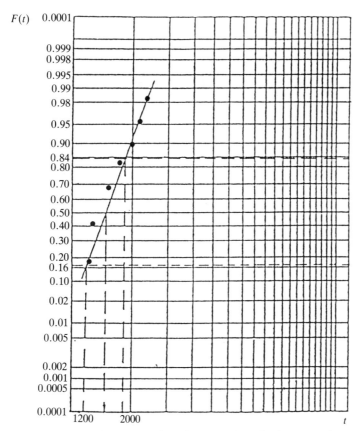

Figure 14.22 Graphical presentation of lognormal distribution, Example 14.6.

Fig. 14.23. After the line of best fit is plotted the parameters which define the distribution can be found by the following procedure:

1. Draw a vertical line down from the point of intersection of the best fit line with ordinate $F(x) = 0.632$, and read the value for A on the horizontal axis, as shown in Fig. 14.24. This holds true because

$$F(A) = 1 - \exp(-(A/(A)^B) = 1 - \exp(-(1)^B) = 0.632$$

2. Draw a line parallel to the line of best fit through the point 0_1 and the value for B can be read on the scale above.

Example 14.7 The results tabulated represent the number of cycles to failure of 11 springs during a laboratory test:

Spring no. :	1	2	3	4	5	6	7	8	9	10	11
No of cycles :	3900	3300	9800	4270	14 000	5640	12 900	13 200	3700	1700	2100

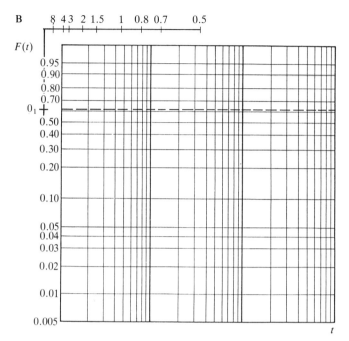

Figure 14.23 Probability paper for the Weibull distribution.

According to experience with springs of a similar design, the operating time to failure can be represented by the Weibull distribution. Check this hypothesis, plot the cumulative function and determine the number of cycles which corresponds to the probability of failure of 0.85.

SOLUTION Applying Eq. (14.2) the numerical values for $F'(t_i)$ were calculated as shown in Table 14.10. In order to apply the graphical method it is necessary to plot the empirical data on Weibull distribution probability paper. Figure 14.25 clearly shows that the plotted points form a straight line, and there is no reason to reject the hypothesis made.

In order to determine the parameters A and B the procedure described above should be applied. From Fig. 14.25 the numerical values obtained are $A = 7700$ and $B = 1.6$. The cumulative distribution function shown in Fig. 14.26 can be used to determine the number of cycles up to which failure will occur, with a probability of 0.85.

If the plotted points on Weibull probability paper form a curve this can mean that we are dealing with a three-parameter Weibull distribution, i.e. that the source parameter C is not zero. The process to determine a numerical value of C is an iterative one which starts with an arbitrarily selected value for C, which obviously must be less than the minimum result obtained. The idea is to try different feasible values for C until something approximating to a straight line is arrived at. If C is underestimated or overestimated there will be a tendency towards curvature in the directions

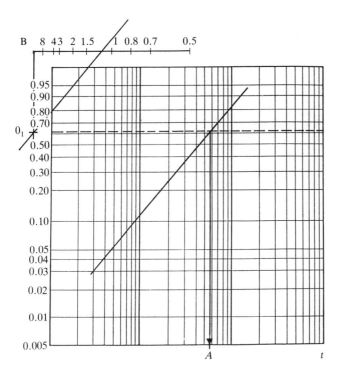

Figure 14.24 Illustration of the graphical method, Weibull distribution.

Table 14.10 Values for Example 14.7

i	t_i	$F'(t_i)$	i	t_i	$F'(t_i)$
1	1700	0.06140	7	5640	0.58772
2	2100	0.14912	8	9800	0.67544
3	3300	0.23684	9	12900	0.76316
4	3700	0.32456	10	13200	0.85088
5	3900	0.41228	11	14000	0.93860
6	4270	0.50000			

indicated by Fig. 14.27, and adjustments should be made in the appropriate direction.

According to some authors the best arbitrary value for the source parameter is $C = 0.9(x_1)$. When the value for C is chosen it is necessary to replot all points again, using the coordinates, $x', F'(x')$ where $x' = x - C$. If the data follow the Weibull distribution the points will lie in a straight line. As explained, this is an iterative process and sometimes it will be necessary to do a few corrections until the best value for C is found. The determination of the other parameters is identical to the case of the

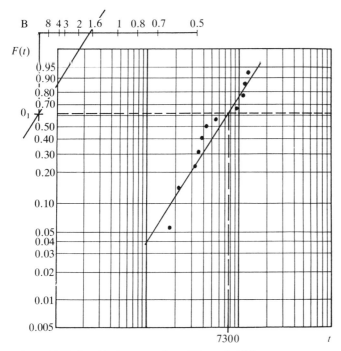

Figure 14.25 Graphical presentation of Example 14.7.

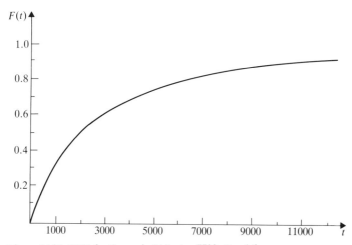

Figure 14.26 *CDF* for Example 14.7, $A = 7700$, $B = 1.6$.

two-parameter distribution. In the case of the three-parameter Weibull distribution an explanation for the existence of the source parameter must be considered.

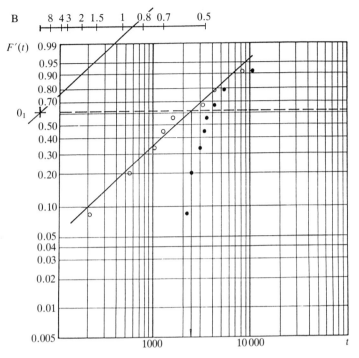

Figure 14.27 Graphical method for determination of parameters, Weibull distribution.

Example 14.8 The empirical data tabulated below represent the time for restoration of the clutch for eight tractors.

Tractor no. :	1	2	3	4	5	6	7	8
Repair time (min) :	330	220	630	300	350	1040	255	520

Determine the probability of replacing the clutch in 300 minutes.

SOLUTION Numerical values of $F'(t_i)$ were determined using the method explained for the Weibull distribution (see Table 14.11) and plotted as shown in Fig. 14.28. It is clear that the points do not follow a straight line, so the hypothesis of a two-parameter Weibull distribution has to be rejected. Assuming that there is a basis for the existence of a third parameter, we will adopt a value of $200 = 0.9(220)$ for the minimum value parameter. The recalculated values of t' are given in the table below:

i :	1	2	3	4	5	6	7	8
t_i :	220	255	300	330	350	520	630	1040
t'_i :	20	55	100	110	250	320	430	840

Points with coordinates $t'_i, F'(t'_i)$ are replotted on the Weibull probability paper as shown in Fig. 14.28. As the new distribution of the points, marked by circles, forms a straight line, the hypothesis of three-parameter Weibull distribution can be accepted. The numerical values of the scale and shape parameters are found in the usual way, thus $A = 448$, $B = 0.9$.

Table 14.11 Values for Example 14.8

i	t_i	$F'(t_i)$	i	t_i	$F'(t_i)$
1	220.0	0.08333	5	350.0	0.55952
2	255.0	0.20238	6	520.0	0.67857
3	300.0	0.32143	7	630.0	0.79762
4	330.0	0.44048	8	1040.0	0.91667

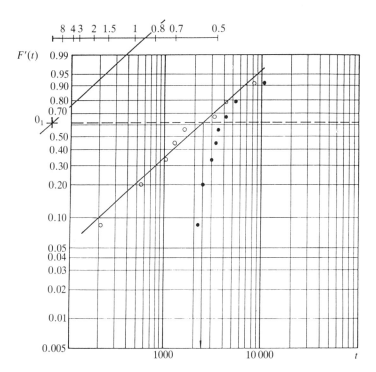

Figure 14.28 Graphical method for the three-parameter Weibull distribution.

The probability of replacing a clutch in 300 minutes can now be calculated, according to Eq. (6.35)

$$P(\text{replacement} < 300) = \exp{-[(300 - 200)/(448 - 200)]^{0.9}} = 0.643$$

Thus, there is a chance of 68 per cent that the clutch will be replaced within 300 hours of maintenance.

14.4 ANALYTICAL DETERMINATION OF DISTRIBUTION FUNCTION PARAMETERS

Regardless of the type of probability distribution involved, the procedure is the same for all cases where the analytical method is used. The main idea of this approach is to fit the linear regression line, $y = ax + b$, through empirical data by using the least squares method. This can be done only after the coordinates related to the numerical values of empirical data and their cumulative probabilities, $x_i, F'(x_i)$ where $i = 1, n$, have been linearized. The parameters a and b of the regression line can be determined according to the following expressions

$$a = \frac{\sum\limits_{i=1}^{n} x_i y_i - \left(\sum\limits_{i=1}^{n} x_i \sum\limits_{i=1}^{n} y_i\right)/n}{\sum\limits_{i=n}^{n} x_i^2 - \left(\sum\limits_{i=1}^{n} x_i\right)^2/n} \tag{14.18}$$

$$b = \sum\limits_{i=1}^{n} y_i/n - a \times \sum\limits_{i=1}^{n} x_i/n \tag{14.19}$$

In order to determine the strength of the relationship between the empirical data and the regression line it is necessary to calculate the coefficient of determination, CD, applying the following expression

$$CD = \frac{\left(\sum\limits_{i=1}^{n} x_i \times y_i - \left(\sum\limits_{i=1}^{n} x_i \times \sum\limits_{i=1}^{n} y_i\right)/n\right)^2}{\left(\sum\limits_{i=1}^{n} x_i^2 - \left(\sum\limits_{i=1}^{n} x_i\right)^2/n\right)\left(\sum\limits_{i=1}^{n} y_i^2 - \left(\sum\limits_{i=1}^{n} y_i\right)^2/n\right)} \tag{14.20}$$

Given that the value obtained for CD is satisfactory it is necessary to find the relationship between parameters a and b of the regression line, and parameters A and B or C of the selected probability distribution. As this relationship is specific to each type of theoretical distribution, the appropriate expressions will be given below and illustrated by numerical examples.

In spite of the fact that this is an analytical method it is still possible to plot all the points defined by the empirical data on the coordinate system with equally spaced axes. Thus, there is no need for the use of special probability paper. As the plotting of data does not cause a great deal of trouble and serves as a good tool for checking the distribution of the points, this exercise is recommended.

For situations where there are 50 or less empirical data values columns 4, 5 and 6 of Table T2 give the numerical values of the coordinates X_i together with values of their sum, $S1$, and the sum of their square values, $S2$. Table T3 should be used if more than 50 data values are available.

14.4.1 Exponential distribution

In order to linearize coordinates, the following expression should be used

$$X_{e_i} - \ln(1 - F'(x_i)) \tag{14.21}$$

$$Y_{e_i} = x_i \tag{14.22}$$

Calculated values for X_{e_i} in the case that the total number of empirical data is less than 50 can be found in Table T2 under the heading $Xe(i)$, otherwise consult Table T3. The numerical value of the parameter A, the single parameter which fully defines the exponential distribution, can be found according to the following expression

$$A = |a| + b \tag{14.23}$$

Example 14.9 The following results represent the support time, in hours, needed for the provision of the specified spare parts for 14 aeroplanes of a particular type and make.

i :	1	2	3	4	5	6	7	8	9	10	11	12	13	14
t_i :	102	209	14	57	54	32	67	134	152	27	230	66	61	34

Check the hypothesis of an exponential distribution using the analytical method and determine the numerical value for A.

SOLUTION The first step towards solving this problem is to classify all results into ascending order (from minimum to maximum value). For the determination of numerical values of X_{e_i} the following two possibilities are available:

1. Use the information given in Table T2, for $n = 14$.
2. Calculate $F'(t_i)$ using Eq. (14.2) and then calculate X_{e_i} using Eq. (14.21).

Regardless of method used, the results should be as shown in Table 14.12. The first step is to evaluate the coefficient of determination by applying Eq. (14.20), thus

$$CD_e = \frac{(-1915.65 - (-13.21 \times 1239)/14)^2}{(22.007 - (174.5/14)(169\,911 - (1\,535\,121/14))} = 0.966$$

According to the high value of CD_e (maximum 1) the hypothesis about an exponential distribution has not been proved incorrect. Hence, the support time for the provision of spares for the aircraft can be modelled by the exponential distribution function.

Numerical values of the parameters a and b can be found by applying Eq. (14.18) and (14.19). Thus, $a = -78.26$ and $b = 14.66$. Consequently, by making use of Eq. (14.23), the value of the scale parameter is $A = 92.9$, which is very close to the value obtained by using the graphical method in Example 14.3.

Graphical representation of this example is shown in Fig. 14.29 where the straight line $Y_e = -78.26 \times X_e + 14.66$ is plotted through the empirical data.

14.4.2 Normal distribution

In the case of the analytical method, polynomial approximation for the determination of numerical values of X_{n_i} has been used by Abramovitz (1964),

Table 14.12 Values for Example 14.9

i	$t_i = Y_{e_i}$	$Y_{e_i}^2$	$F'(t_i)$	X_{e_i}	$X_{e_i}^2$	$Y_{e_i} \times X_{e_i}$
1	14	196	0.04861	-0.0498	0.00248	-0.69720
2	27	729	0.11806	-0.1256	0.01578	-3.39120
3	32	1024	0.18750	-0.2076	0.04310	-6.64320
4	34	1156	0.25694	-0.2970	0.08821	-10.09800
5	54	2916	0.32639	-0.3951	0.15610	-22.52070
6	57	3249	0.39583	-0.5039	0.25392	-28.72230
7	61	3721	0.46528	-0.6260	0.39188	-38.18600
8	66	4356	0.53472	-0.7651	0.58538	-50.49660
9	67	4489	0.60417	-0.9268	0.85896	-62.09560
10	102	10404	0.67361	-1.1197	1.25373	-114.20940
11	134	17956	0.74306	-1.3589	1.84661	-182.09260
12	152	23104	0.81250	-1.6740	2.80228	-254.44800
13	209	43681	0.88194	-2.1366	4.56506	-446.54940
14	230	52900	0.95139	-3.0239	9.14397	-695.49700
	1239	169911	–	-13.21	22.00746	-1915.64720

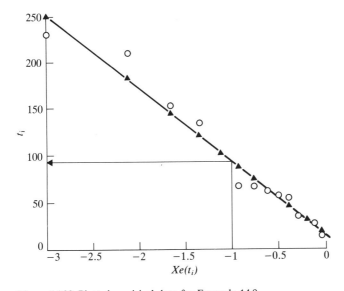

Figure 14.29 Plotted empirical data for Example 14.9.

thus

$$X_{n_i} = \frac{p - c_0 + c_1 \times p + c_2 \times p^2}{1 + d_1 \times p + d_2 \times p^2 + d_3 \times p^3} \qquad (14.24)$$

where

$$p = \sqrt{\ln[1/(1 - F'(x_i)^2]},$$

$$c_0 = 2.515517, \quad c_1 = 0.802853, \quad c_2 = 0.010328$$

$$d_1 = 1.432788, \quad d_2 = 0.189269, \quad d_3 = 0.001308$$

To facilitate calculations employing the above equation column 4 of Table T2 presents numerical values for X_{n_i} where the total number of results is less than 50. (Table T3 should be consulted where there are more than 50 observed values.)

Coordinates on the vertical axis, Y_n, are fully defined by the numerical values of the empirical data, x_i, where $i = 1, n$. The specific member of the family can be found by making use of the following expressions

$$A = b \qquad (14.25)$$

$$B = a \qquad (14.26)$$

Example 14.10 In a laboratory test of 19 washing machines, the operating time to failure of drive belts was recorded. The data obtained (working hours) were:

129	174	118	209	185	98	143
124	164	174	169	110	134	195
140	151	147	156	161		

Assuming that we are dealing with a normal distribution, determine its parameters by applying the analytical method.

SOLUTION In order to determine numerical values for A and B it is necessary to calculate X_{n_i} and Y_{n_i}, $i = 1, 19$. Table T2 can be used, for $n = 19$, or X_{n_i} can be calculated by using Eq. (14.2) where $F'(x_i)$ should be determined by using the median rank method. The data obtained are presented in Table 14.13.

$$CD_n = \frac{(523.92 - 0.22 \times 2886/19)^2}{(15.44 - 0.0484/19)(454\,114 - (2886)/19)} = 0.99$$

As the coefficient of determination is an extremely high value there is no reason to reject the hypothesis of a normal distribution. Parameters A and B can be determined by using the above equations, thus

$$B = a = \frac{523.92 - (0.22 \times 2886)/19}{(15.44 - 0.0484/19)(454\,114 - (2886)/19)} = 31.9$$

$$A = 2886/19 - 31.9 \times (0.22/19) = 151.9$$

Results obtained by the graphical method for the same set of empirical data (see Example 14.4) are very close to those calculated here. A graphical representation of the empirical data analysed and plotted on the ordinary coordinate system is shown in Fig. 14.30.

Table 14.13 Values for Example 14.10

i	$ti = Y_{n_i}$	$Y_{n_i}^2$	$F'(t_i)$	X_{n_i}	$X_{n_i}^2$	$Y_{n_i} \times X_{n_i}$
1	98.0	9604	0.03608	-1.6784	2.82	-164.48
2	110.0	12100	0.08763	-1.3075	1.69	-143.83
3	118.0	13924	0.13918	-1.0590	1.12	-124.96
4	124.0	15376	0.19072	-0.8610	0.74	-106.76
5	129.0	16641	0.24227	-0.6907	0.48	-98.10
6	134.0	17956	0.29381	-0.5374	0.29	-72.01
7	139.0	19321	0.34536	-0.3952	0.16	-54.93
8	143.0	20449	0.39691	-0.2600	0.07	-37.18
9	147.0	21609	0.44845	-0.1291	0.02	-18.98
10	151.0	22801	0.50000	0.0000	0.00	0.00
11	156.0	24336	0.55155	0.1293	0.02	20.17
12	161.0	25921	0.60309	0.2609	0.07	42.00
13	164.0	26896	0.65464	0.3974	0.16	65.17
14	169.0	28561	0.70619	0.5419	0.29	91.58
15	174.0	30276	0.75773	0.6987	0.49	121.57
16	179.0	32041	0.80928	0.8751	0.77	162.77
17	186.0	34596	0.86082	1.0841	1.18	201.74
18	195.0	38025	0.91237	1.3557	1.84	264.36
19	209.0	43681	0.96392	1.7985	3.23	375.89
	2886.0	454114	–	0.2200	15.44	523.92

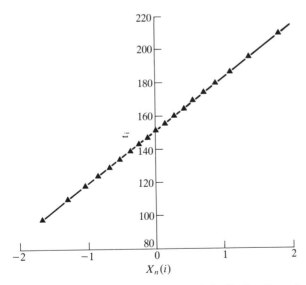

Figure 14.30 Plotted points for the normal distribution, Example 14.10.

14.4.3 Lognormal distribution

The process of determining numerical values of parameters which define the lognormal distribution is similar to the process for the normal distribution, which has already been explained. The linearized values of X_l can be obtained in the same way as Xn, whereas for the values of Y_l the following expression should be used

$$Y_l = \ln(x_i) \tag{14.27}$$

The scale parameter A and shape parameter B can be determined by using the following relationships

$$A = \exp(a) \tag{14.28}$$

$$B = \ln\left(\frac{1}{2}\left(\frac{X_{0.5}}{X_{0.16}} + \frac{X_{0.84}}{X_{0.5}}\right)\right) \tag{14.29}$$

Example 14.11 Five newly-designed engines were driven the equivalent of 30 000 miles. Odometer readings were recorded whenever corrective maintenance action occurred. The following represent odometer readings at which the first failure of particular components occurred.

i :	1	2	3	4	5
t_i :	2600	3800	5100	7700	10 200

(a) Can the data be represented by the lognormal distribution?
(b) If so, estimate the expected operating time of that particular component.

SOLUTION The answer to the first task can be found by determining the numerical value for CD. Making use of Eq. (14.20) and the values established in Table 14.14, it can be found that $CD_l = 0.9918$.

As the answer to part (a) is positive, parameters A and B must be evaluated in order to estimate the expected operating life. This can be done by making use of Eqs (14.28) and (14.29) and the values obtained are $A = 8.546$ and $B = 0.6254$.

The expected operating time to failure can be obtained using Eq. (6.28), thus

$$E(TTF) = MTTF = \exp(A + 0.5B^2) = \exp(8.546 + 0.157) = 6373.9$$

Figure 14.31 shows the empirical data plotted on the ordinary coordinate system.

Table 14.14 Values for Example 14.11

i	t_i	Y_{l_i}	$Y_{l_i}^2$	$F'(t_i)$	X_{l_i}	$X_{l_i}^2$	$Y_{l_i} \times X_{l_i}$
1	2600	7.8632	1.8299	0.12938	-1.1002	1.2104	-8.65109
2	3800	8.2427	67.9421	0.31541	-0.4784	0.2321	-3.94330
3	5100	8.5369	72.8786	0.50000	0.0000	0.0000	0.00000
4	7700	8.9489	80.0828	0.68519	0.4818	0.2321	4.31158
5	10200	9.2301	85.1947	0.87037	1.1282	1.2728	10.41339
	29400	34.8218	367.9281	–	-0.2234	2.9474	-2.13057

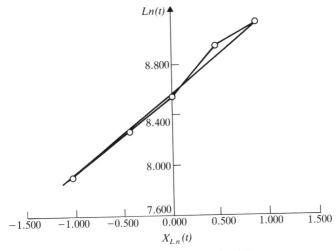

Figure 14.31 Graphical presentation of Example 14.11.

14.4.4 Weibull distribution

In order to determine the parameters which define a Weibull distribution it is necessary to linearize coordinates defined by numerical values of the random variable considered and the corresponding cumulative probability in the following way

$$X_{w_i} = \ln[\ln(1/(1 - F'(x_i)))] \tag{14.30}$$
$$Y_{w_i} = \ln(x_i) \tag{14.31}$$

Calculated values for X_{e_i} are to be found in Table T2 column 6, for the cases where the total number of empirical data is less than 50. Once the parameters a and b for the line of best fit are determined, it is possible to calculate the distribution parameters A and B by using the following equations

$$A = \exp(b) \tag{14.32}$$
$$B = 1/a \tag{14.33}$$

Example 14.12 The results shown below represent the number of cycles to failure of 11 springs during a laboratory test

Spring no. :	1	2	3	4	5	6	7	8	9	10	11
No. of cycles :	3900	3300	9800	4270	14 000	5640	12 900	13 200	3700	1700	2100

Experience with springs of a similar design suggests that the operating time to failure has a Weibull distribution. Check the hypothesis made.

SOLUTION Making use of Table T2 the data presented in Table 14.15 were obtained.
According to Eq. (14.20), the coefficient of determination in this case is $CD_w = 0.912$. Solving this problem by the analytical method the following values were obtained for

Table 14.15 Values for Example 14.12

i	t_i	$Y_{w_i} = \ln(t_i)$	$Y_{w_i}^2$	$F'(t_i)$	X_{w_i}	$X_{w_i}^2$	$X_{w_i} \times Y_{w_i}$
1	1700	7.44	55.33	0.06140	-2.7588	7.61	-20.53
2	2100	7.65	58.52	0.14912	-1.8233	3.32	-13.95
3	3300	8.10	65.64	0.23684	-1.3083	1.71	-10.60
4	3700	8.22	67.50	0.32456	-0.9355	0.88	-7.69
5	3900	8.27	68.37	0.41228	-0.6320	0.40	-5.23
6	4270	8.36	69.88	0.50000	-0.3665	0.13	-3.06
7	5640	8.64	74.61	0.58772	0.1210	0.01	-1.05
8	9800	9.19	84.46	0.67544	0.1180	0.01	-1.08
9	12 900	9.46	89.59	0.76316	0.3649	0.13	3.45
10	13 200	9.49	90.02	0.85088	0.6434	0.41	6.11
11	14 000	9.55	91.14	0.93860	1.0261	1.05	9.80
	–	94.37	815.06	–	-5.7978	15.66	-41.66

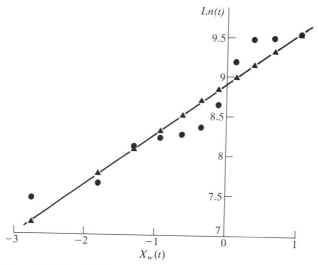

Figure 14.32 Graphical presentation of Example 14.12.

parameters a and b

$$a = \frac{-41.67 - (5.79 \times 94.37)/11}{15.66 - (33.52)/11} = 0.63$$

$$b = 94.37/11 - 0.63(-5.79)/11 = 8.91$$

Therefore, from the above expressions, the scale parameter A and the shape parameter B take the values $A = \exp(8.91) = 7700$ and $B = 1/0.63 = 1.59$, see Fig. 14.32.

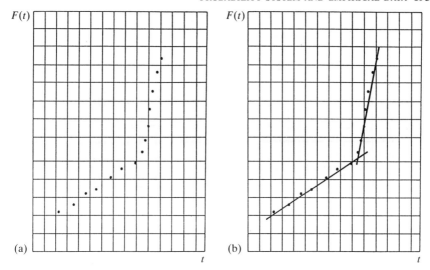

Figure 14.33 Complex distribution.

14.4.5 Complex distribution

In situations where the plotted points on probability paper do not form a straight line, and a straight line cannot be obtained after several trials for the three-parameter distribution, it can mean that we are dealing with a complex distribution. This means that the empirical data consists of two or more subsets of data, related to different variables.

A typical example of a complex distribution for a hypothetical probability is presented in Fig. 14.33(a). In this case all points related to the empirical data must be separated into two or more subsets. Each subset is defined by the points through which a straight line can be fitted (see Fig. 14.33(b)). After the separation the same procedure for determination of the probability distribution, as explained above, must be applied to each set of data.

Example 14.13 In order to determine the number of contacts established by contactors of electrical vehicles, the fleet owner mounted specially designed counters which recorded the activities of contactors over a certain period of days. For maintenance purposes, it is necessary to determine the daily number of contacts.

196	198	199	214	219	226	239	251	263	266	269	273	283	288	292
293	299	315	389	397	412	423	444	451	458	462	466	469	475	479
483	485	488	492	495	499	512	516	527	538	542	551	562	567	580

SOLUTION Applying the method explained, the empirical data are plotted on Weibull paper as shown in Fig. 14.34. It is clear that the distribution of data does not follow a straight line and that it is possible to divide them into 2 subsets, as shown in Fig. 14.35. In order to determine the probability distribution of the first subset of data, the points defined by the following data

196	198	199	214	219	226	239	251	263	269	273	283	288	292	293	299	315

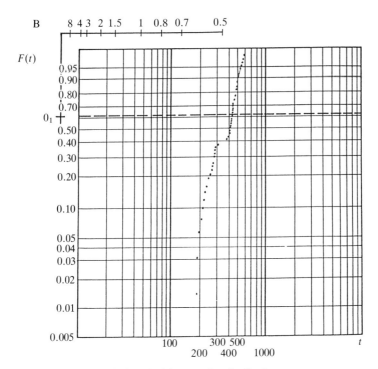

Figure 14.34 Graphical method for complex distribution.

are replotted on separate probability paper and the values obtained are $A = 267.03$, $B = 2.5$ and $C = 166.6$. The same procedure is then repeated for the second subset of data and the calculated values are $A = 506$, $B = 3.7$ and $C = 314.2$.

Analysing the initial empirical data, it was discovered that for several days two vehicles had a considerably longer itinerary than usual which explains the need for the separation of the data.

14.5 HYPOTHESIS TESTING

According to the algorithm given in Fig. 14.1, the last phase of this procedure is to test the hypothesis made. The main objective here is to determine how close the empirical probability distribution is to the postulated theoretical distribution. The hypothesis can be tested by graphical and analytical methods. Regardless of the method used for testing the hypothesis, one of two outputs is possible:

(a) Acceptance with certain confidence, which means that the test performed has not rejected the hypothesis made, and the theoretical distribution

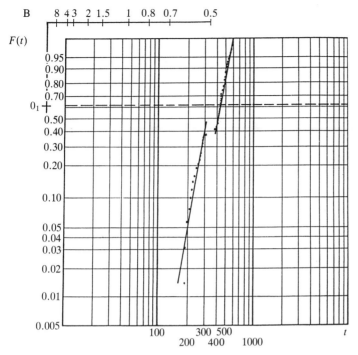

Figure 14.35 Graphical method for determination of Weibull parameters for the first set of data.

function under consideration can be used as the interpretation of the behaviour of the population defined by the random variable.

(b) Rejection with certain probability, which means that there are strong discrepancies between the pattern of distribution of the empirical data analysed and the theoretical distribution considered.

14.5.1 Graphical method

This method for testing the hypothesis is applicable to all cases where the graphical method has been used for determination of the parameters for the theoretical probability distribution.

As has already been mentioned the rank distribution exists for any level of probability. As the median rank is used for the determination of the numerical values of $F'(x_i)$, very often the 5 and 95 per cent rank are used to obtain confidence limits on probability paper (Table T2). By plotting the best line through the points defined by median rank we assume that this line represents the true probability distribution. Any deviation from this straight line can be regarded as an error of that particular data. Thus, any point which does not lie on a straight line can be projected horizontally to it and then the 5 and 95 per cent ranks are related to the straight line.

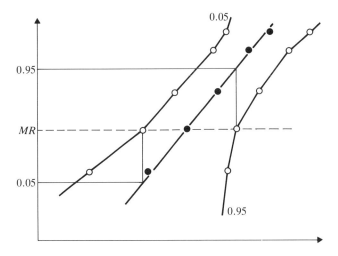

Figure 14.36 Confidence limits.

If the confidence limits are projected about the original sample points, the confidence limits will take an irregular rather than a smooth curve. The area between 5 and 95 per cent curves presents a 90 per cent confidence band about the theoretical distribution function. Thus, if all the data points lie within the envelope formed by the confidence limits then there is no reason for rejecting the distribution as being a reasonable model for the empirical data, Fig. 14.36.

Example 14.14 Let us assume that the empirical data of a random variable considered are as follows:

$$i: \quad 1 \quad 2 \quad 3 \quad 4 \quad 5 \quad 6 \quad 7$$
$$x_i: \quad 75 \quad 82 \quad 93 \quad 99 \quad 102 \quad 121 \quad 137$$

As the main objective of this example is to demonstrate the procedure for hypothesis testing, using the graphical method, the probability paper of some hypothetical probability distribution will be used.

For a sample size 7, values for 50 (median rank), 5 and 95 per cent ranks are obtained from Table T2 as shown in Table 14.16.

Table 14.16 Rank values for sample size 7

$i:$	1	2	3	4	5	6	7
$x_i:$	75	82	93	99	102	21	137
5 :	0.00730	0.05337	0.12876	0.22532	0.34126	0.47830	0.65184
50 :	0.09428	0.22849	0.36412	0.50000	0.63588	0.77151	0.90572
95 :	0.34816	0.52070	0.65874	0.77468	0.87124	0.94662	0.99270

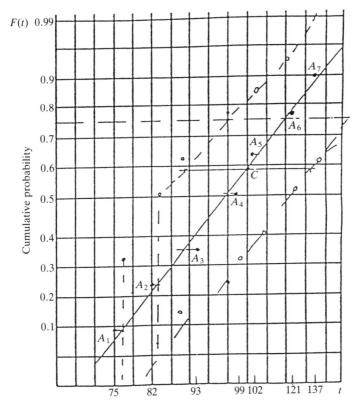

Figure 14.37 Diagrammatical representation of confidence limits.

Point A_1 denotes the plot obtained from the lowest value of random variable X, i.e. x_i, and the corresponding value of MR. Thus point A_1 is fully defined by coordinates 79 and 0.09428. The point A_2 is defined by 82 and 0.22849, and the other 5 points are defined by the values given in Table 14.16. According to the algorithm given in Fig. 14.1 the straight line through the points Ai, $i = 1, 7$, should be fitted next, as shown in Fig. 14.37. The plotted best line misses point A_1. Thus, to plot confidence limits about the theoretical distribution function, point A_1 must be projected horizontally to the best line. This projection defines point $B1$ as shown in Fig. 14.37.

It can be seen from Table 14.16 that the 5 and 95 per cent ranks for the first point are 0.09428 and 0.34816 respectively. These values should be projected vertically to the best line at point $B1$. If the same procedure is repeated for all seven points the confidence band would be obtained (see Fig. 14.37). In order to clarify the obtained confidence interval let us consider point $C = 100$ on the abscissa. It can be said that we are 90 per cent confident that between 34 and 81 per cent of the population will have a value less than or equal to 100. The large width of this interval can be attributed to the small number of results.

The graph can be used as a source for other kinds of information. Namely, if we are interested in the range of values within which, say, 60 per cent of results will have a value with a confidence of 90 per cent, we shall look at the line defined by probability 0.60 on the ordinate. From Fig. 14.37 it can be concluded that the interval in question is defined by 90.5 and 138.5.

14.5.2 Analytical method

One of the most frequently used methods for hypothesis testing is the Kolmogorov–Smirnov test. This involves comparing the empirical cumulative distribution function, $F'(x_i)$ with the theoretical one defined by the distribution parameters. The latter should be calculated for each empirical data point $F(x_i), i = 1, n$. The essence of this test is the determination of the absolute differences (d_i) between the following two cumulative probabilities

$$d_i = \{|F(x_i) - F'(x_i)|, \quad i = 1, n\}$$

as illustrated in Fig. 14.38.

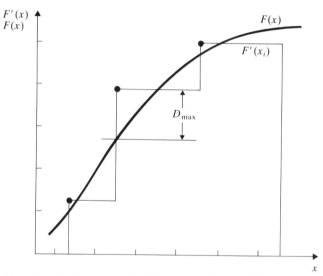

Figure 14.38 Illustration of the Kolmogorov–Smirnov test.

The difference with the highest numerical value should be identified, noted as D_{max} and compared with the value assigned to the level of confidence in percentage terms given in Table T5. If the value obtained for the maximum difference is less than the recommended value p_r then the hypothesis made can be accepted with the chosen confidence level.

$$D_{max} = \{d_i, \quad i = 1, n\}_{max} < p_r$$

SET THEORY

A.1 SETS

Any well-defined collection of objects is a set.

The term well-defined in the above definition means that it must be possible to define the set so that one can decide whether any given object does or does not belong. The individual objects making up the set are known as the *elements* or *members* of the set.

Sets are usually symbolized by capital letters, $A, B, \ldots, Z$, and individual elements by small letters $a, b, c, \ldots, z$. The symbol $\in$ indicates *member of*. Thus, the expression $a \in A$ indicates that element a is the member of set A. For example, all motor vehicles produced by the hypothetical car manufacture company Mobilia (see Table A1.1) in the first three months of 1987 comprise a set, and every particular vehicle is an element of that set. All Profits produced that year make another set, as do all 1.6 litre Profits produced in 1987, or in the summer of 1987, or on 9 October 1987, or any other specified selection, present one set.

Theoretically, every set can be defined in the following two ways:

1. *Listing all of the elements*—for example the expression

$$M = \{\text{Profit, Targa, Cargo}\}$$

   presents a set M whose members are types of the model of Mobilia cars produced in 1987 (see Table A1.1). In practice it means that every type of car produced by Mobilia belongs to the above set.

2. *Specifying the rule which defined the set*—for example:

$$M = \{m : m \text{ is the model of Mobilia produced in 1987}\}$$

Table A1.1 Models of Mobilia cars made in 1987

Engine	Profit	Model Targa	Cargo
1600	X,L,XL,GL	—	—
1800	X,L,XL,GL	—	—
2000	XL,GL	GT	XL,GL
2000 Diesel	XL,GL	—	XL
2300	—	GT	GL
2300 Diesel	—	—	XL
2800	—	GT	GL
2800 Diesel	—	—	GL

The symbol ':' is read 'such that', so the above expression presents the set M of elements m such that m is the model of Mobilia produced in 1987. Given the rule, it is possible to decide immediately whether or not the object belongs to the set. Usually it is more convenient to define a set by its rule than by listing the elements which, in some cases, is almost impossible.

A.2 SUBSETS

Set B is a subset of set A if each element from B is an element of A.

Let us analyse the set S defined as

$$S = \{s : s \text{ is a Profit produced in 1987}\}$$

According to Table A1.1 Profits can be fitted with one of the following engines: 1.6, 1.8, and 2.0. So the set S is split into several groups, for example all Profits produced in 1987 with 1.6 litre engines are denoted $S_{1.6}$, thus

$$S_{1.6} = \{s : s \text{ is a Profit produced in 1987 with 1.6 l engine}\}$$

Other groups can be formed similarly as $S_{1.8}$ and $S_{2.0}$. These groups, in terms of mathematical set theory present subsets of the analysed set. Hence, all Mobilia Profits with 1.6 litre engines produced in 1987 are elements of the set $S_{1.6}$ which is part of the set S. A mathematical description of this relation is:

$$S_{1.6} \subset S$$

It is necessary to emphasize that the opposite is not valid, because it is not true that every Profit produced in that year had a 1.6 litre engine.

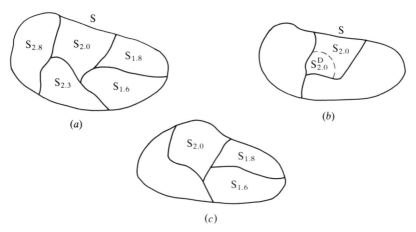

Figure A2.1 Venn diagrams.

If every single element in B is in A, and every single element in A is in B, then the two sets are equivalent or equal, $A = B$. Therefore, two sets are equal if they contain precisely the same elements. If we suppose that in 1987 for some reason Mobilia fitted every Profit with a 1.6 l engine then set $S_{1.6}$ would be equal to the set S.

A very useful method for illustrating sets and for showing relations between them is the well-known Venn diagram which pictures a set as a circle, square, or any other closed geometrical figure which embraces all elements. For example, a Venn diagram picturing set S and its subsets would look like Fig. A2.1(a). Since, say, $S_{1.8}$ is a subset of S, its area is completely included in the area of S. Now let us consider a set whose elements are 2.0 l diesel Profits produced in 1987, denoted as $S_{2.0D}$. As $S_{2.0D}$ is a subset of the set $S_{2.0}$ the Venn diagram would be shown by Fig. A2.1(b). On the other hand, if we take into consideration set $S_{1.8}$ which is not a subset of $S_{2.0}$ then Fig. A2.1(c) would be the Venn diagram because they have no members in common.

Any set that contains no members is called the *empty set*. It is not hard to think of examples; the set of all Mobilias produced in 1762, the set of all Mobilias with two engines, the set of all Mobilias which have won the race in Le Mans. The set of all Profits produced with 1.0 litre engine, $S_{1.0}$, is another example of an empty set which can be described as:

$$S_{1.0} = \emptyset$$

A.3 RELATIONSHIPS BETWEEN SETS

Sometimes it is necessary to *operate* with subsets of a set in order to form new sets, each of which is also a subset of the main set. One of these operations is the *union* of two or more sets. Considering set, say, Z and its two subsets A and B, then by definition

> The union of A and B, written $A \cup B$, is the set of all elements that are members of A, or of B, or of both.
>
> $$A \cup B = \{a : a \in A \text{ or } \in B, \text{ or both}\}$$

For example the union of sets $S_{1.6}$ and $S_{1.8}$, which are both subsets of S is a set of all Profits which are fitted with an engine less than 2.01 capacity, which we shall denote as $S_{2.0}$:

$$S \subset 2.0 = S_{1.6} \cup S_{1.8} = \{x : x \text{ is Mobilia Profit with 1.6 l engine, or}$$
$$x \text{ is Mobilia Profit with 1.8 l engine, or}$$
$$x \text{ is Mobilia Profit with 1.6 l and 1.8 l engine}\}$$

In the Venn diagram, Fig. A3.1, the shaded portion shows the union of $S_{1.6}$ and $S_{1.8}$.

The idea of the union may be extended to more than two sets. In general the union of, say, k sets $A_1, A_2, \ldots, A_k$ is defined as follows:

$$\bigcup_{i=1}^{k} A_i = A_1 \cup A_2 \cup A_3 \cup \ldots \cup A_k$$

A second possible relation between two subsets is the *intersection* of A and B, defined as:

> The intersection of set A and B, written as $A \cap B$, is the set of all members belonging to both A and B
>
> $$A \cap B = \{a : a \in A \text{ and } a \in B\}$$

For example, let us consider universal set M

$$M = \{c : c \text{ is car produced by Mobilia}\}$$
$$S = \{c : c \text{ is engine capacity of Profit}\}$$
$$D = \{c : c \text{ is engine capacity of Targa}\}$$

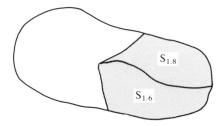

Figure A3.1 Venn diagram showing union.

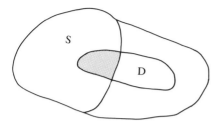

Figure A3.2 Venn diagram showing intersection.

Then the intersection of S and D is set $S \cap D$ written:

$$S \cap D = \{c : c \text{ is engine capacity in Profit and Targa}\}$$

and represents the set of all engines which can be fitted in Profits and Targas. The intersection of the two sets is presented as a Venn diagram in Fig. A3.2.

If the intersection of two sets is empty, then the sets are said to be *mutually exclusive*. Sets $S_{1.6}$ and $S_{1.8}$ are mutually exclusive because their intersection is an empty set. Clearly no Mobilia Profit is produced with both a 1.6 and a 1.8 litre engine, thus

$$S_{1.6} \cap S_{1.8} = \emptyset$$

In general, the intersection of, say, k sets $A_1, A_2, \ldots, A_k$ is defined as follows

$$\bigcap_{i=1}^{k} A_i = A_1 \cup A_2 \cup A_3 \cup \ldots \cup A_k$$

Another operation on sets is called taking the *complement* of a set. Given a universal set, say, Z and its subset A, then the complement of A is A' which is made up of all members of Z that are *not* in A. The union of A and A', however, always equals the universal set Z so that $A \cup A' = Z$. The complement of a set, shown as the shaded area in the Venn diagram of Fig. A3.2, may be expressed:

The complement of $A = A' = \{a : a \text{ which is not a member of } A\}$

The intersection of any set and its complement is always empty, $A \cap A' = \emptyset$, since no element can be a member of both at the same time. For example the complement of set $S_{2.8}$ is set $S'_{2.8}$ which is made up of all elements of set S which are not fitted with 2.8 litre engines.

The *difference* between two sets is closely allied to the idea of a complement. For a given universal set, say Z, the difference between its subsets A and B is a set $A - B = A \cap B'$ which contains all elements that are members of A but *not* members of B. For example, the difference between sets S and C is the set

$$S - C = \{c : c \text{ is Mobilia which is not diesel}\}$$

Table A3.1 Set operations

Universal	Subsets	Operation	Symbol	New set
Z	A, B	Union	$\cup$	$A \cup B$
		Intersection	$\cap$	$A \cap B$
	A	Complement	$'$	$A' = Z - A$
	A, B	Difference	$-$	$A - B$

Although it makes no impact which set we write first, in the symbols for union and intersection, since $A \cup B = B \cup A$ and $A \cap B = B \cap A$, the sequence is very important for the difference between two sets. The difference $A - B$ is *not* the same as the difference $B - A$. Set operations are summarized in Table A3.1.

A.4 MATHEMATICAL RELATIONSHIPS

A mathematical relation on two sets A and B is a subset of $A \times B$

Sometimes it is necessary to define a special kind of set which is made up of a *pair of objects*, symbolized by (a, b), where a is an element of a set A, and b of a set B. The order of listing is significant because of the role which each element plays in the pair. Hence, the pair (a, b) is not necessarily the same as the pair (b, a). The set of all possible pairs (a, b) from A and B is called the *set product* of A and B, $(A \times B)$, thus

$$A \times B = \{(a, b) : a \in A, b \in B\}$$

The product $A \times B$ is one set, whereas the product $B \times A$ would be a different set, with members (b, a). In other words, any mathematical relation is a subset of a set product, but it is necessary to emphasize that it is a specific set of pairs out of all possible pairs.

Being a set, a mathematical relation can be specified in either of the two ways used for defining a set; firstly, by a listing of all pairs that qualify for the relation and, secondly, a statement of the rule by which a pair qualifies, the more usual way of specifying a relation. If S is the set of all Mobilia Profits produced in the UK, and W is the set of all colours, a relation R might be specified:

$$R = \{(s, w) \in (S \times W) : \text{the colour of } s \text{ is } w\}$$

In specifying relations it is necessary to distinguish the different roles that sets of elements play because in a relation which is a subset of $A \times B$, it may be true that only *some* elements a enter into one or more (a, b) pairs.

Hence there is a need to find a way for specifying those elements of A that *are* paired, with b elements, in the relation itself. This subset of elements of A actually figuring in the relation is called the *domain*.

> Given some relation S, which is a subset of $A \times B$:
>
> the domain of $S = \{a \in A : (a,b) \in s$, for at least one $b \in B\}$

The domain of the relation S is always a subset of A and it includes each element in A that actually plays a role in the relation, *vis-a-vis* some b.

Members of the set B that are actually paired with members of A in a specified relation are called the *range*.

> The set of all elements of b in B paired with at least one a in A for the relation S is called the range of S:
>
> range of $S = \{b \in B : (a,b) \in S$, for at least one $a \in A\}$

It is possible for the domain to be equal to A, and for the range to be equal to B, in some examples. In others, the range and the domain will be proper subsets of A and B, respectively.

A.5 FUNCTIONAL RELATIONSHIPS

> A relation where each member of the domain is paired with only one member of the range is a function, or functional relation.

In mathematics and other disciplines there is a special notation for specifying functions and variables. The variable is only a place holder, which can *always* be replaced by any particular element from a set of possibilities. Thus, if X, y, T, v, etc. denote variables, it means that wherever they appear in a mathematical expression they can be replaced by *one* element from some specified set.

It is necessary to emphasize that in spite of its name, a variable is not something that varies, it is a symbol which stands for various values. Another important characteristic is called a *constant*. Unlike a variable, it is a symbol that can be replaced by one and only one number in a given expression and it is a constant for the expression. For example, consider the well-known expression: $A = \pi r^2$. The symbols A and r here are variables; each stands for any one of an infinite set of positive numbers, whereas the symbol π is a constant, which can be replaced only with a particular number. Regardless of the values used for r and A, π is always the same.

Variables are the basis for the notation most commonly used for functions. Given the variables v and s, each with a specified range, a relation between the variables is a subset of all possible (a, b) pairs. If this relation is a function, then this fact is symbolized by any of the statements such as $v = f(s)$, $v = G(s)$, and so on. In other words, these expressions all state that there is a rule which pairs each possible value of s with at most one value of v. The different Roman and Greek letters that precede the symbol (x) simply indicate different rules, giving different functions or subsets of all (a, b) pairs. The notation does nothing more than indicate that the function rule exists.

A symbol $f(s)$ actually stands for a value of v; this is the value of v that is paired with s by the function rule. The symbol $f(v)$ is *not* the function, but an indicator that the function rule exists, turning a value s into some value $f(s)$ or v. Sometimes the function rule itself is stated: $v = f(s) = 3s + 5$, $v = G(s) = \exp(t/s)$, $s > 0$, and so on. The function rule stated permits the mapping of values of s to corresponding values of v.

Clearly, the idea of a function can be extended to include any number of variables, for example the well-known function for calculating engine capacity

$$EC = f(b, s, c)$$

where b is the bore, s the stroke and c is the number of cylinders and all variables ranging over real numbers. The rule which assigns values to the EC, when b, s and c are known, is:

$$EC = \frac{\pi}{4} \times \text{bore}^2 \times \text{stroke} \times \text{number of cylinders}$$

$$= \frac{\pi}{4} \times b^2 \times s \times c$$

A.6 SEQUENCES OF EVENTS

Suppose that any trial of an experiment must result in one of k mutually exclusive and exhaustive events, $(A_1, \ldots, A_k)$. If the experiment is repeated n times, an elementary event may be the set of outcomes as a series of observations. This leads to a sequence of events: the outcome of the first trial, the outcome of the second, and so on in order through to the outcome of the nth trial. The outcome of the whole series of trials might be the sequence $(A_3, A_1, A_2, \ldots, A_3)$. This denotes that event A_3 occurred on the first trial, A_1 on the second trial, A_2 on the third, and so on. The place in order gives the trial on which the event occurred, and the symbol occupying that place shows the event occurring for that trial.

The number of different sequences that a given series of trials can produce is provided by the following six rules.

Rule 1

If any one of k mutually exclusive and exhaustive events can occur on each of n trials, then there are k^n different sequences that may result from a set of trials.

As an example of this rule, consider an MOT test for cars. Each test can result only in a pass (P) or a failed (F) event, thus, $k = 2$. If the car is tested three years consecutively ($n = 3$) the total number of possible results of testing the car is $k^n = 2^3 = 8$ sequences. Exactly the same number is obtained for possible outcomes of testing three cars simultaneously. Possible sequences are

PPP, FFF, PFF, PPF, FPF, FFP, FFP, PFP

Rule 2

If $k_1, k_2, \ldots, k_n$ are the numbers of distinct events that can occur on trials $1, 2, \ldots, n$ in a series, then the number of different sequences of n events that can occur is

$$k_1 \times k_2 \times \ldots \times k_n$$

How many different 7-place number plates are possible if the first three places are to be occupied by letters, followed by three places of numbers and finishing with a letter? The answer by direct application of rule 2 is:

$$26 \times 26 \times 26 \times 10 \times 10 \times 10 \times 26 = 456\,976\,000$$

It is not difficult to see that in the case that the same number of events can occur on any trial, say k, the total number of sequences is k multiplied by itself n times, or k^n. So, rule 1 is actually a special case of rule 2.

Rule 3

The number of different ways that n distinct things may be arranged in order is:

$$n! = 1 \times 2 \times 3 \times \ldots \times (n-1) \times n$$

where $0! = 1$ and $1! = 1$. An arrangement in order is called a permutation, so the total number of permutations of n objects is $n!$, where the symbol $n!$ is called 'n-factorial'.

As an illustration of this rule, let us analyse a taxi-cab company which has ten cars and ten drivers. How many ways could the drivers be assigned to the cars?

Any of the drivers could be put into the first car, making ten possibilities for car 1. But, given the occupancy of car 1, there are only nine drivers for car 2; the total number of ways cars 1 and 2 could be filled is $10 \times 9 = 90$. Now, we shall consider car 3. With cars 1 and 2 occupied, eight drivers remain to be allocated, so that there are $10 \times 9 \times 8 = 720$ ways to fill cars 1, 2, and 3. Finally, when nine cars have been filled there remains only one driver to fill the remaining place, so there are:

$$10 \times 9 \times 8 \times 7 \times 6 \times 5 \times 4 \times 3 \times 2 \times 1 = 10! = 3\,628\,800$$

ways of arranging the ten drivers into the ten cars.

Rule 4

The number of ways of selecting and arranging r objects from among n distinct objects is:

$$\frac{n!}{(n-r)!}$$

As an example, let us consider a taxi-cab company again, but this time, we shall imagine that there are only five cars. How many different ways could the manager select ten drivers and arrange them into the available cars?

There are ten ways that the first car might be filled, nine ways for the second, and so on, until car 5 could be filled in six ways. Thus there are $10 \times 9 \times 8 \times 7 \times 6$ ways to select drivers to fill the five cars. This number is equivalent to $(10!/5!) = (n!/(n-r)!)$, the number of ways that ten drivers out of ten may be selected and arranged, but divided by the number of arrangements of the five unselected drivers in the five *missing* cars.

Rule 5

The total number of ways of selecting r distinct combinations of n objects, irrespective of order, is:

$$\frac{n!}{r!(n-r)!} = \binom{n}{r}$$

The symbol $\binom{n}{r}$ is *not* a symbol for a fraction, it denotes the number of combinations of n things, taken r at a time. Sometimes $\binom{n}{r}$ is replaced by the symbols nCr or rCn.

As an example of the use of this rule, suppose that a total of seven vehicles are repaired in a garage. Four of them can be repaired simultaneously. How many ways could they be selected if they arrived in the garage at the same time?

Here, $r = 4$, $n = 7$ so that:

$$\binom{n}{r} = \frac{7!}{4! \times 3!} = \frac{1 \times 2 \times 3 \times 4 \times 5 \times 6 \times 7}{(1 \times 2 \times 3 \times 4)(1 \times 2 \times 3)} = 35$$

Rule 5 can be interpreted as the total number of ways of dividing a set of n objects into two subsets of r and $(n-r)$ objects, since the determination of the first subset of r objects unequally determines the second subset of $(n - r)$ objects, which consists of the objects remaining after the first subset has been selected. This can be generalized to the situation in which the set of n objects is to be divided into k subsets, where $k > 2$:

Rule 6

The total number of ways of dividing a set of n objects into k mutually exclusive and exhaustive subsets of $N_1, N_2, \ldots, N_k$ objects, respectively, where $N_1 + N_2 + \ldots + N_k = n$, is:

$$\frac{n!}{N_1! \times N_2! \times \ldots \times N_k!}$$

Rule 5 is a special case of rule 6, with $K = 2$, $N_1 = r$, and $N_2 = N - r$.

As an example of Rule 6, consider an extension of the previous example. Suppose a total of nine vehicles have to be serviced at a specific garage which operates in three shifts. Four of them are repaired in the morning shift, three in the afternoon and two in the night shift. Here $n = 9$, $k = 3$, $N_1 = 4$, $N_2 = 3$, and $N_3 = 2$, so that the number of combinations is

$$\frac{n!}{N_1! N_2! N_3!} = \frac{9!}{4!\, 3!\, 2!} = \frac{(1 \times 2 \times \ldots \times 8 \times 9)}{(1 \times 2 \times 3 \times 4)(1 \times 2 \times 3)(1 \times 2)} = 2520$$

ProbChar USER GUIDE

B.1 ProbChar MENU MAP

MAIN MENU

1) Input Menu
2) Empirical Characteristics Menu
3) Determination of Probability Distribution
4) Probability Characteristics Menu
5) Stated Value Menu
6) Quit the Program

STATED VALUE MENU

1) Probability Functions at a Stated Value
2) Value at a Stated Probability
3) Probability of a Stated Interval
4) Return to the Main Menu

PROBABILITY CHARACTERISTICS MENU

1) Plot the Cumulative Distribution Function
2) Plot the Probability Density Function
3) Summary of the Characteristics
4) View the Data
5) Return to the Main Menu

DISTRIBUTION MENU

1) Plot the Normal Paper
2) Plot the LogNormal Paper
3) Plot the Exponential Paper
4) Plot the Weibull Paper

5) Distribution Comparison
6) View the Data
7) Return to the Main Menu

EMPIRICAL CHARACTERISTICS MENU

1) Plot the Cumulative Distribution Function
2) Plot the Probability Density Function
3) View the Data
4) Summary of the Empirical Data
5) Proceed with the Calculations
6) Return to the Input Menu

INPUT MENU

1) Enter Classified Data from the Keyboard
2) Enter Unclassified Data from the Keyboard
3) Enter Classified Data from a File
4) Enter Unclassified Data from a File
5) Add to Unclassified Data on File
6) Enter a Probability Distribution
7) Return to the Main Menu

B.2 INTRODUCTION TO ProbChar

ProbChar is a software package which provides:

- Selection of the theoretical best fit probability distribution.
- Estimation of the best fit parameters (scale, shape and location).
- Testing by the Kolmogorov–Smirnov test.
- Calculation of the main probability functions.
- Determination of summary measures.

for the random variable of the empirical data under consideration. The random variable could be related to any process, experiment, observation or test which has been repeated several times, and for which three or more different results have been obtained.

ProbChar also offers the following features :

- An easy to use menu system.
- Data can be entered in several different ways.
- Collected empirical values can be in different forms.
- The empirical data can be edited.
- A theoretical probability distribution can be used.
- Both empirical and estimated characteristics are available.

- Estimated probability characteristics can be calculated for user-specified values of the random variable.
- All the functions can be represented graphically.
- Probability paper is displayed on the screen together with the data points, the theoretical best fit line and confidence limits.
- Both text and graphical results can be printed as a hard copy on a dot matrix or laser printer.
- Graphics screens can be dumped to a disk in both PCX and TIFF formats.

B.3 GENERAL POINTS ON USING ProbChar

B.3.1 Note about compatibility

This software will work on any IBM PC or compatible computer with at least 512K of RAM, and running under DOS version 3.0 or later. However, the graphics features only function with EGA, VGA or Hercules graphics monitors. Text output is always available. A hard copy of both text and graphical output can be made if your computer has either an Epson-compatible or LaserJet II-compatible parallel printer connected. ProbChar may be run from either a floppy or hard disk, but hard disk operation is recommended.

B.3.2 Characteristics ProbChar will calculate

This software package calculates the following probability characteristics :

- Expected value of the random variable
- Standard deviation of the random variable
- Variability of the random variable
- Cumulative distribution function
- Probability density function

The data needed for the calculations are the available numerical values of the random variable under consideration, which can be measured in any units (hours, cycles, pounds, kilograms, metres, watts, litres, etc.).

B.3.3 Getting started

To run ProbChar the computer has to be running under MS-DOS or PC-DOS, so if it has just been turned on it must be 'booted' from a system disk (or hard disk). To run from a floppy disk, insert the ProbChar disk into a disk drive and type 'probchar' to start the program. On hard disk the program may be in a separate directory, and in this case simply change to this directory and type 'probchar' to start the program.

When the software is first run a title screen will appear. If ProbChar detects a printer then you will be asked to select which type you have connected (see Sec. B.5). Then ProbChar will offer the option:

Press H for the Help Pages   Press RETURN to Continue

Pressing H (or h) will provide information about the program and how to use it. The help pages are essentially a summary of the information given in this appendix. To scroll up and down through the help pages use the PgUp and PgDn keys on the keyboard numeric keypad. If the keys appear not to work then make sure that the Num Lock is OFF. When the RETURN key is pressed the program will offer to print out the help pages — if a printer is detected — or proceed to the Input Menu. The help pages are available from any menu by pressing F1, and the first help page displayed will be relevant to that menu. Pressing RETURN will either move back to the present menu or the option to print.

B.3.4 Using the menus

ProbChar uses a system of menus to enable you to select the required options. The menu screens consist of a list of numbered options. The title of the menu is at the top of the screen and the option currently ready for selection is highlighted and indicated by arrows. There are two ways to choose an option from the menu. If the required option is highlighted then pressing the RETURN key selects it. If the desired option is not highlighted, then the up and down cursor keys or a mouse (if one is connected and its driver is installed) will move the highlight bar, and arrows, up and down the list of options; pressing RETURN as before will select that option. Alternatively, pressing the number key corresponding to the desired option will immediately select it.

There is a main menu from which five sub-menus can be selected. The last option of each of the sub-menus returns you to the main menu, and the last option of the main menu will quit the program and return you to DOS. The sub-menus available are :

1) Input Menu
2) Empirical Characteristics Menu
3) Distribution Menu
4) Probability Characteristics Menu
5) Stated Value Menu

Information reminding you of what to do is displayed on the bottom line of the screen.

B.4 DEALING WITH RAW DATA

The Input Menu can be selected from the Main Menu at any time. Select the appropriate option according to how and where your data is stored. If you select the wrong option at the Input Menu then by pressing RETURN without entering any characters into the filename you will move straight back to the menu. However, any current data will have been lost, and so it will have to be re-loaded from the data disk.

After leaving the title page, the first menu that you will see is the Input Menu. The data is entered after choosing the appropriate option from this menu. The data required is information on the random variable whose probability characteristics are to be calculated. The larger the volume of data entered, the more accurate the results should be. As mentioned before, the random variable can be measured in a variety of units. To ensure that the program can display the information clearly, the values for the random variable should not exceed 1 000 000. For example if time is measured in minutes and the maximum value is 1 200 000 then the data should be converted to hours, so the maximum value becomes 20 000 hours. The program will not accept any negative values or values above 1 000 000, and if it reads any out of range values from a file then it will truncate them so as to fall within the random variable range of 0 to 1 000 000.

Empirical values can be entered from the keyboard or loaded from a disk. Data entered from the keyboard will automatically be saved on disk if there are three or more random variable values. If there are ever less than three data values at the end of an input procedure then the data file is erased from the disk.

With classified data, only files that have been previously created by ProbChar can be loaded. However, with unclassified data the file may be created by ProbChar or any text editor which does not add control codes and stores it as an ASCII text file. If you use a text editor then all the entries should be placed flush with the left-hand edge of the page, and only one value should be entered per line. The name of the random variable (e.g. 'Time to Failure of the Fan Belts') should be entered in the first line of the file (the name can be up to 36 characters long), and the units of the random variable should be entered on the next line (e.g. 'miles travelled' — the units can consist of up to 30 characters). If the top line is left blank then the program assigns the name 'random variable' to any output, and similarly if the second line is left blank the program assigns 'Units' to the random variable. The data filename should not contain an extension (e.g. '.dat'), and must be a valid filename under DOS. If your text editor adds any extension to the filename then the file should be renamed so that it has no extension.

B.4.1 Entering data from keyboard

If you want to enter data from the keyboard — either as a new file or adding into an unclassified data file — then you should choose the appropriate option from the Input Menu. You will then be prompted to enter the filename. If you are entering data into a new file and ProbChar finds a data file with the same name then you will be asked if you want to overwrite the existing file. If you reply 'Yes' then the existing file is erased. If you reply 'No' then you will be returned to the Input Menu and the disk file will remain intact. If you are adding data into an existing unclassified file and ProbChar cannot find the file then an error message is displayed and you are returned to the Input Menu.

If the filename that you enter is valid then you will be prompted to enter the name of the random variable and its units. You can enter your own parameters, but if you press RETURN, without entering any response, at either of the requests then the default reply is chosen for that request, 'random variable' and 'Units' appropriately. These parameters are written to the data file and are used throughout the program for the presentation of data and results.

To add your data values, or interval frequencies, to the list displayed on the right-hand side of the screen, simply enter them when prompted; this is entry mode. By pressing RETURN without entering any value you will proceed to the data viewing section of ProbChar — 'Table of Empirical Data'.

When entering data from the keyboard it is also possible to edit the data: edit mode. The data is listed on the right-hand side of the screen, with a pair of pointers (> <) to indicate the current data value. To begin edit mode enter E (or e) at the entry mode prompt, and to return to entry mode press RETURN at the edit mode prompt without entering a value. To scroll the list by one value use the up- and down-arrow keys on the numeric keypad, and to scroll ten values at a time use PgUp and PgDn — it is not possible to scroll off either end of the list.

To edit a value you must be in edit mode. Then scroll the list so that the value to be edited is between the pointers, and press E (or e) and you will be prompted to enter the new value. Press RETURN to enter the value into the list. If you are at the 'Table of Empirical Data' then the empirical data will be recalculated and displayed on the left-hand side of the screen. You can move between entry and edit mode as many times as you like, so long as there is data to enter (see the paragraph below).

If you are entering classified data, then when you leave edit mode and return to entry mode, the data list will scroll so that the first interval which has not been edited or had a data value entered into it will be ready to accept the next value. If no such interval exists then you will move straight to the 'Table of Empirical Data' after the data has been saved to disk. If you are entering unclassified data, then when you return to entry mode the next entered value will be inserted after the current value (as indicated by the pointers), as will all the following values. If you are at the 'Table of Empirical Data' then you can only operate in edit mode, and pressing RETURN at the prompt ends the procedure.

If you are adding data to an unclassified file (option 5 of the Input Menu) then, as well as being able to add items to the list, you can edit all of the data.

If you are entering classified data then it is not possible to have zero frequency in the first data interval. If you find that you have no data in the first interval then you must shift all your interval boundaries upwards and reclassify your data. If you are entering classified data and you want to quit (but don't mind erasing the data file), then enter a zero for the first interval and press RETURN to finish editing/entering the data — the data file will be erased and the data will be lost.

When you have finished entering and editing the data then press RETURN and the random variable name and units and the data values will be saved onto the disk under the filename entered earlier. You can now review the data (see Sec. B.4.3 below).

To determine which mode you are operating in — entry or edit — simply look at the message line at the bottom of the screen. If you have the option to scroll the list using the cursor keys then you are in edit mode, whereas if you are prompted to enter a value then you are in entry mode. When you are entering data from the keyboard the title line (at the top of the screen) will read something like 'Enter xxx Data from the Keyboard', whereas if you are reviewing the data the title line will read 'Table of Empirical Data'.

B.4.2 Data stored on disk

The only difference between loading classified and unclassified data from a file on a disk is the option selected: option 3 for classified data and option 4 for unclassified data. Once you have chosen your option you will be asked for the name of the file. If the file is on another disk then insert that disk into the drive and enter the filename. Providing the file can be found it will be loaded. If your computer has a second disk drive then that cannot be used for data files — you must first copy them onto the ProbChar disk, or swap disks when loading the appropriate files. Data loaded from a disk can still be viewed and edited (see Sec. B.4.3 below).

Please note that if you take the ProbChar disk out of the drive to load a data file from a separate disk then there is no need to replace the ProbChar disk afterwards — ProbChar is self-sufficient once it is loaded — though you will need to keep a disk in the drive if you wish to save and load further data files.

B.4.3 Reviewing the data

After loading a data file or saving entered data, ProbChar will re-display the values so that they can be checked for errors: edit mode. To edit the data use the procedure described in Sec. B.4.1. To finish checking and editing the data press RETURN. If any changes have been made to the data then you will be asked if you would like to save the amended list. If you reply 'No' then the amended data list is still available for the calculations and the disk file will remain untouched. However, if you reply 'Yes' then the new data list is written to disk. You will then be asked if there is a value below which no data can possibly occur. If you are not certain about a value then press N (or n), otherwise press Y (or y) and enter the value (e.g. if you were analysing the ages of UK car drivers then you would reply 'Yes' followed by '17' because the minimum legal age for driving a car in the UK is 17 years old, and so this must be a set minimum value for the data).

B.4.4 Data as a probability distribution

If the probability distribution of the data is already known, then this can be entered using option 6 of the Input Menu. You will be prompted to select one of four probability distributions :

1) Normal Distribution
2) Lognormal Distribution
3) Exponential Distribution
4) Weibull Distribution

Once you have selected the distribution, you will be asked for the appropriate parameters, including the minimum value parameter (if you have selected the lognormal or Weibull distributions). If you are unsure of a value for the minimum value parameter then enter 0 (though this will affect the probability characteristics).

ProbChar will perform a few calculations as you enter the parameters, and if it finds that any of the results will be out of range (below zero or above one million) then an error message is displayed and your data will be lost. ProbChar will also check the value of any minimum value parameter to determine if it is too high, and you may have to re-enter a lower value if your value is rejected at the prompt (and check your data).

Note that if the data is entered in this form then only the Probability Characteristics Menu and Stated Value Menu options are applicable, and so the other sub-menus will not be available.

B.4.5 Classification of unclassified data

If more than 50 items of unclassified data are entered from the keyboard or from a file then the data will be automatically classified. The results will be displayed as for classified data, though the data will be saved as being unclassified, and the original unclassified data can still be viewed and printed when appropriate.

B.5 USING ProbChar

B.5.1 Starting the calculations

Once new data has been entered, either as classified or unclassified data or as a known distribution, then the estimated probability characteristics have to be calculated. There are two ways to start these calculations:

- Select the final option from the Input Menu
- Select the penultimate option from the Empirical Characteristics Menu

The Empirical Characteristics Menu cannot be accessed if your data is in the form of a known distribution, since no empirical data exists.

As these calculations can take a few seconds a 'wait' message appears on the screen during this time. Once the calculations have been performed it will not be necessary to repeat them until new data — in whatever form — is loaded into ProbChar.

B.5.2 Printing and dumping graphics and text

When ProbChar is started, if it detects a parallel printer it will prompt you to enter the type of printer — either 'D' for dot-matrix or 'L' for laser. The dot-matrix printer must be Epson-compatible, and the laser printer must be LaserJet II-compatible. Consult your printer manual for a list of emulations. This option will only be offered once, and so the printer must be switched on and be on-line when ProbChar is begun. ProbChar will assume that the printer's current page is blank, and, in the case of dot matrix printers, that the print head is at the top of the page.

When either text or graphical data is displayed, and a parallel printer is detected, ProbChar will offer the option to print the data. You will be presented with these options when appropriate. Since printing a graphics screen is time consuming, you will be asked to confirm your choice. All the details are presented on the bottom line of the screen.

Graphical data may be dumped to disk in two proprietary formats, PCX and TIFF. Again, you will be presented with these options via the bottom line of the screen, when appropriate. If an error occurs while a graphics file is being written to disk, i.e. the disk becomes full, ProbChar will indicate this on the bottom line of the screen, and the graphics file will be erased. Both formats are suitable for importing into various software, and as such may be included in reports, presentations, etc.

The convention for naming the graphics files is as follows. For known distribution data, the first group of letters indicates the type of distribution, i.e. 'normal', 'lognor', 'expon' or 'weibul'. The next letter will be 'c' to indicate calculated (estimated) data, followed by either 'c' or 'p' to indicate the *CDF* or *PDF* respectively. The file extension will be either 'PCX' or 'TIF' depending upon which format was selected by the user.

For unclassified or classified data, the original data filename provides all but the final two letters of the graphics filename (the data filename may have to be truncated). Two types of graph may be dumped to disk — probability paper or *CDF/PDF*. If the screen dump is of a probability paper then the final two letters will either be 'no', 'lo', 'ex' or 'we' as appropriate to indicate the type of paper. If the screen dump is a *CDF/PDF* plot then the penultimate letter indicates the type of graph — 'c' or 'e' to represent calculated (estimated) or empirical data, followed by either 'c' or 'p' to indicate the *CDF* or *PDF* respectively. The file extension will be either 'PCX' or 'TIF', as before, to indicate the format requested by the user.

B.5.3 Accessing the information available

Written and graphical presentation (where possible) of empirical, distribution or probability data can be viewed by selecting the appropriate option from the Main Menu. For example, to see the Weibull paper plot of the random variable, select option 3 from the Main Menu, and then choose option 4 from the Distribution Menu. This will draw the Weibull probability paper with the data points, line of best fit and confidence limits (if the data is considered to be unclassified). Press RETURN to get back to the Distribution Menu. If no further similar operations are required then select option 7 to return to the Main Menu.

To quit ProbChar, and return to DOS, select option 6 from the Main Menu. To obtain help on the current menu options press F1.

B.5.4 Stated value calculations

By selecting option 5 from the Main Menu — Stated Value Menu — the following estimated characteristics can be obtained:

- Probability density function and cumulative distribution function at a specific value of the random variable.
- Value of the random variable at a certain cumulative probability.
- Probability of the random variable falling within a stated interval.

B.5.5 Probability characteristics menu

By selecting option 4 from the Main Menu — Probability Characteristics Menu — the following functions are available:

- View the graph of the cumulative distribution function, $F(x)$.
- View the graph of the probability density function, $f(x)$.
- See the best probability distribution fit and various items of data relating to that distribution, e.g. the results of the Kolmogorov–Smirnov test.
- Scroll through the tabular listing of the cumulative distribution function and probability density function data.

B.5.6 Distribution menu

The Distribution Menu is selected by choosing option 3 at the Main Menu. The options available enable you to view the empirical data drawn on normal, lognormal, exponential and Weibull probability paper, along with the best fit line for that particular distribution. The axes are labelled and scaled, and the 5 per cent and 95 per cent confidence limits are drawn on if the data is unclassified (or hasn't been classified). A comparison between the four probability distributions is also available (option 5), and the parameters for each distribution are listed along with the coefficients of determination. A tabular listing of the empirical data, converted to normal, lognormal, exponential and Weibull plotting values can also be viewed (option 6).

B.5.7 Empirical characteristics menu

The following options are available from the Empirical Characteristics Menu:

- View the empirical cumulative distribution function, $F'(x)$.
- View the empirical probability density function, $f'(x)$.
- Scroll through a tabular listing of the empirical data values with their corresponding values of $F'(x)$, $f'(x)$ and the 5 per cent and 95 per cent confidence limits (where appropriate).
- Scroll through a list of the empirical data and see the calculated empirical data properties, e.g. the standard deviation.

B.5.8 Seeing the results

Two sets of results are available from ProbChar — the empirical and the estimated (calculated) values. When option 6 is selected at the Input Menu (Enter a Probability Distribution) only the estimated probability characteristics and the stated value options are available. The necessary calculations are initiated by selecting option 7 at the Input Menu.

If, however, ProbChar is supplied with empirical data (classified or unclassified), then both sets of results are available. The empirical results can be viewed, without having to wait for the estimated results to be calculated, by selecting option 7 — Empirical Characteristics Menu at the Input Menu. These results are also still accessible from the Main Menu by selecting option 2. To view the estimated results the appropriate calculations have to be performed. These calculations are initiated by either selecting option 8 at the Input Menu or option 5 from the Empirical Characteristics Menu. The estimated results are presented by the Probability Characteristics Menu and the Distribution Menu (options 3 and 4 of the Main Menu), with the Stated Value Menu (option 5 at the Main Menu) providing further calculated information.

Enquiries concerning problems with the software should be addressed to:

Dr J. Knezevic,
Research Centre, MIRCE,
School of Engineering,
University of Exeter,
Exeter EX4 4QF,
UK.

Table T1 *PMF* and *CDF* for standardized normal variable

z	$F(z)$	$f(z)$	z	$F(z)$	$f(z)$
-4.00	0.00003	0.00013	-3.51	0.00022	0.00084
-3.99	0.00003	0.00014	-3.50	0.00023	0.00087
-3.98	0.00003	0.00014	-3.49	0.00024	0.00090
-3.97	0.00004	0.00015	-3.48	0.00025	0.00094
-3.96	0.00004	0.00016	-3.47	0.00026	0.00097
-3.95	0.00004	0.00016	-3.46	0.00027	0.00100
-3.94	0.00004	0.00017	-3.45	0.00028	0.00104
-3.93	0.00004	0.00018	-3.44	0.00029	0.00107
-3.92	0.00004	0.00018	-3.43	0.00030	0.00111
-3.91	0.00005	0.00019	-3.42	0.00031	0.00115
-3.90	0.00005	0.00020	-3.41	0.00032	0.00119
-3.89	0.00005	0.00021	-3.40	0.00034	0.00123
-3.88	0.00005	0.00021	-3.39	0.00035	0.00127
-3.87	0.00005	0.00022	-3.38	0.00036	0.00132
-3.86	0.00006	0.00023	-3.37	0.00038	0.00136
-3.85	0.00006	0.00024	-3.36	0.00039	0.00141
-3.84	0.00006	0.00025	-3.35	0.00040	0.00146
-3.83	0.00006	0.00026	-3.34	0.00042	0.00151
-3.82	0.00007	0.00027	-3.33	0.00043	0.00156
-3.81	0.00007	0.00028	-3.32	0.00045	0.00161
-3.80	0.00007	0.00029	-3.31	0.00047	0.00167
-3.79	0.00008	0.00030	-3.30	0.00048	0.00172
-3.78	0.00008	0.00031	-3.29	0.00050	0.00178
-3.77	0.00008	0.00033	-3.28	0.00052	0.00184
-3.76	0.00008	0.00034	-3.27	0.00054	0.00190
-3.75	0.00009	0.00035	-3.26	0.00056	0.00196
-3.74	0.00009	0.00037	-3.25	0.00058	0.00203
-3.73	0.00010	0.00038	-3.24	0.00060	0.00210
-3.72	0.00010	0.00039	-3.23	0.00062	0.00216
-3.71	0.00010	0.00041	-3.22	0.00064	0.00224
-3.70	0.00011	0.00042	-3.21	0.00066	0.00231
-3.69	0.00011	0.00044	-3.20	0.00069	0.00238
-3.68	0.00012	0.00046	-3.19	0.00071	0.00246
-3.67	0.00012	0.00047	-3.18	0.00074	0.00254
-3.66	0.00013	0.00049	-3.17	0.00076	0.00262
-3.65	0.00013	0.00051	-3.16	0.00079	0.00271
-3.64	0.00014	0.00053	-3.15	0.00082	0.00279
-3.63	0.00014	0.00055	-3.14	0.00084	0.00288
-3.62	0.00015	0.00057	-3.13	0.00087	0.00297
-3.61	0.00015	0.00059	-3.12	0.00090	0.00307
-3.60	0.00016	0.00061	-3.11	0.00094	0.00317
-3.59	0.00017	0.00063	-3.10	0.00097	0.00327
-3.58	0.00017	0.00066	-3.09	0.00100	0.00337
-3.57	0.00018	0.00068	-3.08	0.00103	0.00347
-3.56	0.00019	0.00071	-3.07	0.00107	0.00358
-3.55	0.00019	0.00073	-3.06	0.00111	0.00369
-3.54	0.00020	0.00076	-3.05	0.00114	0.00381
-3.53	0.00021	0.00079	-3.04	0.00118	0.00393
-3.52	0.00022	0.00081	-3.03	0.00122	0.00405

Table T1 *cont.*

z	$F(z)$	$f(z)$	z	$F(z)$	$f(z)$
−3.02	0.00126	0.00417	−2.53	0.00570	0.01625
−3.01	0.00131	0.00430	−2.52	0.00587	0.01667
−3.00	0.00135	0.00443	−2.51	0.00604	0.01709
−2.99	0.00139	0.00457	−2.50	0.00621	0.01752
−2.98	0.00144	0.00470	−2.49	0.00639	0.01797
−2.97	0.00149	0.00485	−2.48	0.00657	0.01842
−2.96	0.00154	0.00499	−2.47	0.00675	0.01888
−2.95	0.00159	0.00514	−2.46	0.00695	0.01935
−2.94	0.00164	0.00530	−2.45	0.00714	0.01983
−2.93	0.00169	0.00545	−2.44	0.00734	0.02032
−2.92	0.00175	0.00561	−2.43	0.00755	0.02082
−2.91	0.00181	0.00578	−2.42	0.00776	0.02134
−2.90	0.00187	0.00595	−2.41	0.00797	0.02186
−2.89	0.00193	0.00613	−2.40	0.00820	0.02239
−2.88	0.00199	0.00631	−2.39	0.00842	0.02293
−2.87	0.00205	0.00649	−2.38	0.00865	0.02349
−2.86	0.00212	0.00668	−2.37	0.00889	0.02405
−2.85	0.00219	0.00687	−2.36	0.00914	0.02463
−2.84	0.00226	0.00707	−2.35	0.00938	0.02521
−2.83	0.00233	0.00727	−2.34	0.00964	0.02581
−2.82	0.00240	0.00748	−2.33	0.00990	0.02642
−2.81	0.00248	0.00769	−2.32	0.01017	0.02704
−2.80	0.00255	0.00791	−2.31	0.01044	0.02768
−2.79	0.00264	0.00814	−2.30	0.01072	0.02832
−2.78	0.00272	0.00837	−2.29	0.01101	0.02898
−2.77	0.00280	0.00860	−2.28	0.01130	0.02965
−2.76	0.00289	0.00884	−2.27	0.01160	0.03033
−2.75	0.00298	0.00909	−2.26	0.01191	0.03103
−2.74	0.00307	0.00934	−2.25	0.01222	0.03173
−2.73	0.00317	0.00960	−2.24	0.01254	0.03245
−2.72	0.00326	0.00987	−2.23	0.01287	0.03319
−2.71	0.00336	0.01014	−2.22	0.01321	0.03393
−2.70	0.00347	0.01042	−2.21	0.01355	0.03469
−2.69	0.00357	0.01070	−2.20	0.01390	0.03547
−2.68	0.00368	0.01099	−2.19	0.01426	0.03625
−2.67	0.00379	0.01129	−2.18	0.01463	0.03705
−2.66	0.00391	0.01160	−2.17	0.01500	0.03787
−2.65	0.00402	0.01191	−2.16	0.01538	0.03870
−2.64	0.00414	0.01223	−2.15	0.01577	0.03954
−2.63	0.00427	0.01256	−2.14	0.01617	0.04040
−2.62	0.00440	0.01289	−2.13	0.01658	0.04127
−2.61	0.00453	0.01323	−2.12	0.01700	0.04216
−2.60	0.00466	0.01358	−2.11	0.01743	0.04306
−2.59	0.00480	0.01394	−2.10	0.01786	0.04397
−2.58	0.00494	0.01430	−2.09	0.01830	0.04490
−2.57	0.00508	0.01468	−2.08	0.01876	0.04585
−2.56	0.00523	0.01506	−2.07	0.01922	0.04681
−2.55	0.00539	0.01545	−2.06	0.01969	0.04779
−2.54	0.00554	0.01584	−2.05	0.02018	0.04878

Table T1 *cont.*

z	$F(z)$	$f(z)$	z	$F(z)$	$f(z)$
2.04	0.02067	0.04979	−1.55	0.06056	0.11999
−2.03	0.02117	0.05081	−1.54	0.06177	0.12185
−2.02	0.02169	0.05185	−1.53	0.06299	0.12374
−2.01	0.02221	0.05291	−1.52	0.06424	0.12564
−2.00	0.02274	0.05398	−1.51	0.06551	0.12756
−1.99	0.02329	0.05507	−1.50	0.06679	0.12949
−1.98	0.02385	0.05617	−1.49	0.06810	0.13144
−1.97	0.02441	0.05729	−1.48	0.06942	0.13341
−1.96	0.02499	0.05843	−1.47	0.07077	0.13539
−1.95	0.02558	0.05958	−1.46	0.07213	0.13739
−1.94	0.02618	0.06075	−1.45	0.07351	0.13940
−1.93	0.02680	0.06194	−1.44	0.07492	0.14143
−1.92	0.02742	0.06314	−1.43	0.07634	0.14348
−1.91	0.02806	0.06436	−1.42	0.07779	0.14554
−1.90	0.02871	0.06560	−1.41	0.07925	0.14761
−1.89	0.02937	0.06686	−1.40	0.08074	0.14970
−1.88	0.03005	0.06813	−1.39	0.08225	0.15180
−1.87	0.03073	0.06942	−1.38	0.08378	0.15392
−1.86	0.03144	0.07073	−1.37	0.08533	0.15605
−1.85	0.03215	0.07205	−1.36	0.08690	0.15819
−1.84	0.03288	0.07339	−1.35	0.08849	0.16035
−1.83	0.03362	0.07475	−1.34	0.09010	0.16252
−1.82	0.03437	0.07613	−1.33	0.09174	0.16471
−1.81	0.03514	0.07752	−1.32	0.09340	0.16691
−1.80	0.03592	0.07893	−1.31	0.09508	0.16912
−1.79	0.03672	0.08036	−1.30	0.09678	0.17134
−1.78	0.03753	0.08181	−1.29	0.09850	0.17357
−1.77	0.03835	0.08328	−1.28	0.10025	0.17581
−1.76	0.03919	0.08476	−1.27	0.10202	0.17807
−1.75	0.04005	0.08626	−1.26	0.10381	0.18034
−1.74	0.04092	0.08778	−1.25	0.10563	0.18262
−1.73	0.04181	0.08931	−1.24	0.10747	0.18490
−1.72	0.04271	0.09087	−1.23	0.10933	0.18720
−1.71	0.04362	0.09244	−1.22	0.11121	0.18951
−1.70	0.04456	0.09403	−1.21	0.11312	0.19183
−1.69	0.04550	0.09564	−1.20	0.11505	0.19415
−1.68	0.04647	0.09726	−1.19	0.11700	0.19649
−1.67	0.04745	0.09891	−1.18	0.11898	0.19883
−1.66	0.04845	0.10057	−1.17	0.12098	0.20118
−1.65	0.04946	0.10224	−1.16	0.12300	0.20353
−1.64	0.05049	0.10394	−1.15	0.12505	0.20590
−1.63	0.05154	0.10565	−1.14	0.12712	0.20827
−1.62	0.05260	0.10738	−1.13	0.12921	0.21065
−1.61	0.05369	0.10913	−1.12	0.13133	0.21303
−1.60	0.05479	0.11090	−1.11	0.13347	0.21542
−1.59	0.05591	0.11268	−1.10	0.13564	0.21781
−1.58	0.05704	0.11448	−1.09	0.13783	0.22021
−1.57	0.05819	0.11630	−1.08	0.14004	0.22261
−1.56	0.05937	0.11813	−1.07	0.14228	0.22502

Table T1 *cont.*

z	$F(z)$	$f(z)$	z	$F(z)$	$f(z)$
-1.06	0.14454	0.22743	-0.57	0.28429	0.33907
-1.05	0.14683	0.22984	-0.56	0.28769	0.34099
-1.04	0.14914	0.23226	-0.55	0.29111	0.34289
-1.03	0.15147	0.23467	-0.54	0.29455	0.34477
-1.02	0.15383	0.23709	-0.53	0.29800	0.34662
-1.01	0.15622	0.23951	-0.52	0.30148	0.34844
-1.00	0.15862	0.24193	-0.51	0.30497	0.35024
-0.99	0.16106	0.24435	-0.50	0.30848	0.35201
-0.98	0.16351	0.24677	-0.49	0.31201	0.35376
-0.97	0.16599	0.24919	-0.48	0.31556	0.35548
-0.96	0.16849	0.25160	-0.47	0.31912	0.35717
-0.95	0.17102	0.25402	-0.46	0.32270	0.35884
-0.94	0.17357	0.25643	-0.45	0.32630	0.36048
-0.93	0.17615	0.25884	-0.44	0.32991	0.36208
-0.92	0.17875	0.26124	-0.43	0.33354	0.36366
-0.91	0.18138	0.26364	-0.42	0.33718	0.36521
-0.90	0.18402	0.26604	-0.41	0.34084	0.36673
-0.89	0.18670	0.26843	-0.40	0.34452	0.36822
-0.88	0.18939	0.27082	-0.39	0.34821	0.36968
-0.87	0.19211	0.27320	-0.38	0.35191	0.37110
-0.86	0.19486	0.27557	-0.37	0.35563	0.37250
-0.85	0.19762	0.27794	-0.36	0.35936	0.37386
-0.84	0.20042	0.28030	-0.35	0.36311	0.37519
-0.83	0.20323	0.28265	-0.34	0.36686	0.37649
-0.82	0.20607	0.28499	-0.33	0.37064	0.37775
-0.81	0.20893	0.28732	-0.32	0.37442	0.37898
-0.80	0.21182	0.28964	-0.31	0.37822	0.38017
-0.79	0.21472	0.29196	-0.30	0.38202	0.38134
-0.78	0.21765	0.29426	-0.29	0.38584	0.38246
-0.77	0.22061	0.29655	-0.28	0.38967	0.38355
-0.76	0.22359	0.29882	-0.27	0.39351	0.38461
-0.75	0.22658	0.30109	-0.26	0.39736	0.38563
-0.74	0.22961	0.30334	-0.25	0.40123	0.38662
-0.73	0.23265	0.30558	-0.24	0.40510	0.38757
-0.72	0.23572	0.30780	-0.23	0.40898	0.38848
-0.71	0.23881	0.31001	-0.22	0.41287	0.38935
-0.70	0.24192	0.31221	-0.21	0.41676	0.39019
-0.69	0.24505	0.31438	-0.20	0.42067	0.39099
-0.68	0.24821	0.31654	-0.19	0.42458	0.39175
-0.67	0.25138	0.31869	-0.18	0.42850	0.39248
-0.66	0.25458	0.32081	-0.17	0.43243	0.39317
-0.65	0.25780	0.32292	-0.16	0.43637	0.39382
-0.64	0.26104	0.32501	-0.15	0.44031	0.39443
-0.63	0.26430	0.32708	-0.14	0.44426	0.39500
-0.62	0.26758	0.32913	-0.13	0.44821	0.39554
-0.61	0.27088	0.33116	-0.12	0.45217	0.39603
-0.60	0.27420	0.33317	-0.11	0.45613	0.39649
-0.59	0.27754	0.33516	-0.10	0.46010	0.39690
-0.58	0.28091	0.33713	-0.09	0.46407	0.39728

Table T1 *cont.*

z	F(z)	f(z)	z	F(z)	f(z)
−0.08	0.46804	0.39762	0.41	0.65912	0.36675
−0.07	0.47202	0.39792	0.42	0.66278	0.36523
−0.06	0.47600	0.39818	0.43	0.66642	0.36368
−0.05	0.47998	0.39839	0.44	0.67005	0.36210
−0.04	0.48397	0.39857	0.45	0.67367	0.36049
−0.03	0.48795	0.39871	0.46	0.67726	0.35886
−0.02	0.49194	0.39881	0.47	0.68084	0.35719
−0.01	0.49593	0.39887	0.48	0.68441	0.35550
0.00	0.50000	0.39889	0.49	0.68795	0.35378
0.01	0.50403	0.39887	0.50	0.69148	0.35203
0.02	0.50802	0.39881	0.51	0.69499	0.35026
0.03	0.51200	0.39872	0.52	0.69849	0.34846
0.04	0.51599	0.39858	0.53	0.70196	0.34664
0.05	0.51998	0.39840	0.54	0.70542	0.34479
0.06	0.52396	0.39818	0.55	0.70886	0.34291
0.07	0.52794	0.39792	0.56	0.71228	0.34101
0.08	0.53192	0.39762	0.57	0.71568	0.33909
0.09	0.53589	0.39728	0.58	0.71906	0.33715
0.10	0.53986	0.39691	0.59	0.72242	0.33518
0.11	0.54383	0.39649	0.60	0.72576	0.33319
0.12	0.54779	0.39603	0.61	0.72908	0.33119
0.13	0.55175	0.39554	0.62	0.73239	0.32915
0.14	0.55570	0.39501	0.63	0.73567	0.32710
0.15	0.55965	0.39443	0.64	0.73893	0.32503
0.16	0.56359	0.39382	0.65	0.74217	0.32294
0.17	0.56753	0.39318	0.66	0.74539	0.32084
0.18	0.57146	0.39249	0.67	0.74858	0.31871
0.19	0.57538	0.39176	0.68	0.75176	0.31657
0.20	0.57929	0.39100	0.69	0.75492	0.31441
0.21	0.58320	0.39020	0.70	0.75805	0.31223
0.22	0.58709	0.38936	0.71	0.76116	0.31003
0.23	0.59098	0.38849	0.72	0.76425	0.30783
0.24	0.59486	0.38757	0.73	0.76732	0.30560
0.25	0.59873	0.38663	0.74	0.77036	0.30336
0.26	0.60260	0.38564	0.75	0.77338	0.30111
0.27	0.60645	0.38462	0.76	0.77638	0.29885
0.28	0.61029	0.38357	0.77	0.77936	0.29657
0.29	0.61412	0.38247	0.78	0.78232	0.29428
0.30	0.61794	0.38135	0.79	0.78525	0.29198
0.31	0.62175	0.38019	0.80	0.78816	0.28967
0.32	0.62554	0.37899	0.81	0.79104	0.28735
0.33	0.62933	0.37776	0.82	0.79390	0.28501
0.34	0.63310	0.37650	0.83	0.79674	0.28267
0.35	0.63686	0.37520	0.84	0.79956	0.28032
0.36	0.64060	0.37387	0.85	0.80235	0.27796
0.37	0.64433	0.37251	0.86	0.80511	0.27560
0.38	0.64805	0.37112	0.87	0.80786	0.27322
0.39	0.65175	0.36969	0.88	0.81058	0.27084
0.40	0.65544	0.36823	0.89	0.81328	0.26846

Table T1 *cont.*

z	F(z)	f(z)	z	F(z)	f(z)
0.90	0.81595	0.26607	1.39	0.91774	0.15182
0.91	0.81860	0.26367	1.40	0.91925	0.14972
0.92	0.82122	0.26127	1.41	0.92073	0.14763
0.93	0.82382	0.25886	1.42	0.92220	0.14556
0.94	0.82640	0.25645	1.43	0.92364	0.14350
0.95	0.82895	0.25404	1.44	0.92507	0.14145
0.96	0.83148	0.25163	1.45	0.92647	0.13942
0.97	0.83398	0.24921	1.46	0.92786	0.13741
0.98	0.83646	0.24679	1.47	0.92922	0.13541
0.99	0.83892	0.24437	1.48	0.93056	0.13343
1.00	0.84135	0.24195	1.49	0.93189	0.13146
1.01	0.84376	0.23953	1.50	0.93319	0.12951
1.02	0.84614	0.23712	1.51	0.93448	0.12758
1.03	0.84850	0.23470	1.52	0.93575	0.12566
1.04	0.85084	0.23228	1.53	0.93699	0.12376
1.05	0.85315	0.22987	1.54	0.93822	0.12187
1.06	0.85543	0.22745	1.55	0.93943	0.12000
1.07	0.85770	0.22505	1.56	0.94062	0.11815
1.08	0.85993	0.22264	1.57	0.94179	0.11632
1.09	0.86215	0.22024	1.58	0.94295	0.11450
1.10	0.86434	0.21784	1.59	0.94408	0.11270
1.11	0.86651	0.21544	1.60	0.94520	0.11092
1.12	0.86865	0.21306	1.61	0.94630	0.10915
1.13	0.87077	0.21067	1.62	0.94738	0.10740
1.14	0.87286	0.20830	1.63	0.94845	0.10567
1.15	0.87493	0.20592	1.64	0.94950	0.10396
1.16	0.87698	0.20356	1.65	0.95053	0.10226
1.17	0.87900	0.20120	1.66	0.95154	0.10058
1.18	0.88100	0.19885	1.67	0.95254	0.09892
1.19	0.88298	0.19651	1.68	0.95352	0.09728
1.20	0.88493	0.19417	1.69	0.95449	0.09565
1.21	0.88686	0.19185	1.70	0.95543	0.09405
1.22	0.88877	0.18953	1.71	0.95637	0.09246
1.23	0.89065	0.18722	1.72	0.95728	0.09088
1.24	0.89252	0.18493	1.73	0.95819	0.08933
1.25	0.89435	0.18264	1.74	0.95907	0.08779
1.26	0.89617	0.18036	1.75	0.95994	0.08627
1.27	0.89796	0.17809	1.76	0.96080	0.08477
1.28	0.89973	0.17584	1.77	0.96164	0.08329
1.29	0.90148	0.17359	1.78	0.96246	0.08183
1.30	0.90320	0.17136	1.79	0.96327	0.08038
1.31	0.90490	0.16914	1.80	0.96407	0.07895
1.32	0.90658	0.16693	1.81	0.96485	0.07754
1.33	0.90824	0.16473	1.82	0.96562	0.07614
1.34	0.90988	0.16255	1.83	0.96638	0.07476
1.35	0.91149	0.16038	1.84	0.96712	0.07341
1.36	0.91309	0.15822	1.85	0.96784	0.07206
1.37	0.91466	0.15607	1.86	0.96856	0.07074
1.38	0.91621	0.15394	1.87	0.96926	0.06943

Table T1 *cont.*

z	F(z)	f(z)	z	F(z)	f(z)
1.88	0.96995	0.06814	2.37	0.99111	0.02406
1.89	0.97062	0.06687	2.38	0.99134	0.02349
1.90	0.97128	0.06561	2.39	0.99158	0.02294
1.91	0.97193	0.06438	2.40	0.99180	0.02240
1.92	0.97257	0.06316	2.41	0.99202	0.02186
1.93	0.97320	0.06195	2.42	0.99224	0.02134
1.94	0.97381	0.06076	2.43	0.99245	0.02083
1.95	0.97441	0.05959	2.44	0.99266	0.02033
1.96	0.97500	0.05844	2.45	0.99286	0.01984
1.97	0.97558	0.05730	2.46	0.99305	0.01936
1.98	0.97615	0.05618	2.47	0.99324	0.01889
1.99	0.97670	0.05508	2.48	0.99343	0.01842
2.00	0.97725	0.05399	2.49	0.99361	0.01797
2.01	0.97778	0.05292	2.50	0.99379	0.01753
2.02	0.97831	0.05186	2.51	0.99396	0.01710
2.03	0.97882	0.05082	2.52	0.99413	0.01667
2.04	0.97932	0.04980	2.53	0.99430	0.01626
2.05	0.97982	0.04879	2.54	0.99446	0.01585
2.06	0.98030	0.04780	2.55	0.99461	0.01545
2.07	0.98077	0.04682	2.56	0.99477	0.01506
2.08	0.98124	0.04586	2.57	0.99491	0.01468
2.09	0.98169	0.04491	2.58	0.99506	0.01431
2.10	0.98214	0.04398	2.59	0.99520	0.01394
2.11	0.98257	0.04307	2.60	0.99534	0.01358
2.12	0.98300	0.04217	2.61	0.99547	0.01323
2.13	0.98341	0.04128	2.62	0.99560	0.01289
2.14	0.98382	0.04041	2.63	0.99573	0.01256
2.15	0.98422	0.03955	2.64	0.99585	0.01223
2.16	0.98461	0.03871	2.65	0.99598	0.01191
2.17	0.98500	0.03788	2.66	0.99609	0.01160
2.18	0.98537	0.03706	2.67	0.99621	0.01130
2.19	0.98574	0.03626	2.68	0.99632	0.01100
2.20	0.98610	0.03547	2.69	0.99643	0.01071
2.21	0.98645	0.03470	2.70	0.99653	0.01042
2.22	0.98679	0.03394	2.71	0.99664	0.01014
2.23	0.98713	0.03319	2.72	0.99674	0.00987
2.24	0.98745	0.03246	2.73	0.99683	0.00961
2.25	0.98778	0.03174	2.74	0.99693	0.00935
2.26	0.98809	0.03103	2.75	0.99702	0.00909
2.27	0.98840	0.03034	2.76	0.99711	0.00885
2.28	0.98870	0.02966	2.77	0.99720	0.00861
2.29	0.98899	0.02899	2.78	0.99728	0.00837
2.30	0.98928	0.02833	2.79	0.99736	0.00814
2.31	0.98956	0.02768	2.80	0.99744	0.00792
2.32	0.98983	0.02705	2.81	0.99752	0.00770
2.33	0.99010	0.02643	2.82	0.99760	0.00748
2.34	0.99036	0.02582	2.83	0.99767	0.00728
2.35	0.99061	0.02522	2.84	0.99774	0.00707
2.36	0.99086	0.02463	2.85	0.99781	0.00687

Table T1 *cont.*

z	F(z)	f(z)	z	F(z)	f(z)
2.86	0.99788	0.00668	3.35	0.99960	0.00146
2.87	0.99795	0.00649	3.36	0.99961	0.00141
2.88	0.99801	0.00631	3.37	0.99962	0.00136
2.89	0.99807	0.00613	3.38	0.99964	0.00132
2.90	0.99813	0.00595	3.39	0.99965	0.00127
2.91	0.99819	0.00578	3.40	0.99966	0.00123
2.92	0.99825	0.00562	3.41	0.99968	0.00119
2.93	0.99830	0.00545	3.42	0.99969	0.00115
2.94	0.99836	0.00530	3.43	0.99970	0.00111
2.95	0.99841	0.00514	3.44	0.99971	0.00107
2.96	0.99846	0.00499	3.45	0.99972	0.00104
2.97	0.99851	0.00485	3.46	0.99973	0.00100
2.98	0.99856	0.00471	3.47	0.99974	0.00097
2.99	0.99860	0.00457	3.48	0.99975	0.00094
3.00	0.99865	0.00443	3.49	0.99976	0.00090
3.01	0.99869	0.00430	3.50	0.99977	0.00087
3.02	0.99874	0.00417	3.51	0.99978	0.00084
3.03	0.99878	0.00405	3.52	0.99978	0.00081
3.04	0.99882	0.00393	3.53	0.99979	0.00079
3.05	0.99886	0.00381	3.54	0.99980	0.00076
3.06	0.99889	0.00370	3.55	0.99981	0.00073
3.07	0.99893	0.00358	3.56	0.99981	0.00071
3.08	0.99896	0.00348	3.57	0.99982	0.00068
3.09	0.99900	0.00337	3.58	0.99983	0.00066
3.10	0.99903	0.00327	3.59	0.99983	0.00063
3.11	0.99906	0.00317	3.60	0.99984	0.00061
3.12	0.99910	0.00307	3.61	0.99985	0.00059
3.13	0.99913	0.00298	3.62	0.99985	0.00057
3.14	0.99916	0.00288	3.63	0.99986	0.00055
3.15	0.99918	0.00279	3.64	0.99986	0.00053
3.16	0.99921	0.00271	3.65	0.99987	0.00051
3.17	0.99924	0.00262	3.66	0.99987	0.00049
3.18	0.99926	0.00254	3.67	0.99988	0.00047
3.19	0.99929	0.00246	3.68	0.99988	0.00046
3.20	0.99931	0.00238	3.69	0.99989	0.00044
3.21	0.99934	0.00231	3.70	0.99989	0.00042
3.22	0.99936	0.00224	3.71	0.99990	0.00041
3.23	0.99938	0.00217	3.72	0.99990	0.00039
3.24	0.99940	0.00210	3.73	0.99990	0.00038
3.25	0.99942	0.00203	3.74	0.99991	0.00037
3.26	0.99944	0.00196	3.75	0.99991	0.00035
3.27	0.99946	0.00190	3.76	0.99991	0.00034
3.28	0.99948	0.00184	3.77	0.99992	0.00033
3.29	0.99950	0.00178	3.78	0.99992	0.00032
3.30	0.99952	0.00172	3.79	0.99992	0.00030
3.31	0.99953	0.00167	3.80	0.99993	0.00029
3.32	0.99955	0.00161	3.81	0.99993	0.00028
3.33	0.99957	0.00156	3.82	0.99993	0.00027
3.34	0.99958	0.00151	3.83	0.99994	0.00026

Table T1 *cont.*

z	F(z)	f(z)	z	F(z)	f(z)
3.84	0.99994	0.00025	3.93	0.99996	0.00018
3.85	0.99994	0.00024	3.94	0.99996	0.00017
3.86	0.99994	0.00023	3.95	0.99996	0.00016
3.87	0.99995	0.00022	3.96	0.99996	0.00016
3.88	0.99995	0.00021	3.97	0.99996	0.00015
3.89	0.99995	0.00021	3.98	0.99997	0.00014
3.90	0.99995	0.00020	3.99	0.99997	0.00014
3.91	0.99995	0.00019	4.00	0.99997	0.00013
3.92	0.99996	0.00018			

Table T2 Median, 5 and 95 per cent rank

i	$F'(i)$	$1 - F'(i)$		Xn(i)	Xe(i)	Xw(i)	p0.05	p0.95
Total number of results: 1								
1	0.50000	0.50000		0.0000	−0.6931	−0.3665	0.05000	0.95000
			S3:	0.0000	−0.6931	−0.3665		
			S4:	0.0000	0.4804	0.1343		
Total number of results: 2								
1	0.29167	0.70833		−0.5436	−0.3448	−1.0647	0.02532	0.77639
2	0.70833	0.29167		0.5481	−1.2321	0.2088	0.22361	0.97468
			S3:	0.0045	−1.5769	−0.8559		
			S4:	0.5959	1.6371	1.1771		
Total number of results: 3								
1	0.20588	0.79412		−0.8087	−0.2305	−1.4674	0.01695	0.63160
2	0.50000	0.50000		0.0000	−0.6931	−0.3665	0.13535	0.86465
3	0.79412	0.20588		0.8206	−1.5805	0.4577	0.36840	0.98305
			S3:	0.0119	−2.5041	−1.3762		
			S4:	1.3273	3.0314	2.4971		
Total number of results: 4								
1	0.15909	0.84091		−0.9782	−0.1733	−1.7529	0.01274	0.52713
2	0.38636	0.61364		−0.2872	−0.4884	−0.7167	0.09761	0.75139
3	0.61364	0.38636		0.2884	−0.9510	−0.0503	0.24860	0.90239
4	0.84091	0.15909		0.9982	−1.8383	0.6088	0.47237	0.98726
			S3:	0.0211	−3.4508	−1.9110		
			S4:	2.1189	4.5521	3.9595		
Total number of results: 5								
1	0.12963	0.87037		−1.1002	−0.1388	−1.9745	0.01021	0.45072
2	0.31481	0.68519		−0.4784	−0.3781	−0.9727	0.07644	0.65741
3	0.50000	0.50000		0.0000	−0.6931	−0.3665	0.18925	0.81075
4	0.68519	0.31481		0.4818	−1.1558	0.1448	0.34259	0.92356
5	0.87037	0.12963		1.1282	−2.0431	0.7145	0.54928	0.98979
			S3:	0.0314	−4.4088	−2.4544		
			S4:	2.9443	6.1526	5.5103		
Total number of results: 6								
1	0.10938	0.89063		−1.1940	−0.1158	−2.1556	0.00851	0.39304
2	0.26563	0.73438		−0.6196	−0.3087	−1.1753	0.06285	0.58180
3	0.42188	0.57813		−0.1962	−0.5480	−0.6015	0.15316	0.72866
4	0.57813	0.42188		0.1967	−0.8630	−0.1473	0.27134	0.84684
5	0.73438	0.26563		0.6258	−1.3257	0.2819	0.41820	0.93715
6	0.89063	0.10938		1.2300	−2.2130	0.7943	0.60696	0.99149
			S3:	0.0427	−5.3742	−3.0034		
			S4:	3.7911	7.8085	7.1219		
Total number of results: 7								
1	0.09459	0.90541		−1.2694	−0.0994	−2.3089	0.00730	0.34816
2	0.22973	0.77027		−0.7303	−0.2610	−1.3432	0.05337	0.52070
3	0.36486	0.63514		−0.3434	−0.4539	−0.7898	0.12876	0.65874
4	0.50000	0.50000		0.0000	−0.6931	−0.3665	0.22532	0.77468
5	0.63514	0.36486		0.3450	−1.0082	0.0082	0.34126	0.87124
6	0.77027	0.22973		0.7395	−1.4709	0.3858	0.47930	0.94662

Table T2 *Cont.*

i	F'(i)	1 – F'(i)		Xn(i)	Xe(i)	Xw(i)	p0.05	p0.95
7	0.90541	0.09459		1.3132	− 2.3582	0.8579	0.65184	0.99270
			S3:	0.0546	− 6.3446	− 3.5565		
			S4:	4.6528	9.5053	8.7781		
Total number of results: 8								
1	0.08333	0.91667		− 1.3319	− 0.0870	− 2.4417	0.00639	0.31234
2	0.20238	0.79762		− 0.8206	− 0.2261	− 1.4867	0.04639	0.47068
3	0.32143	0.67857		− 0.4601	− 0.3878	− 0.9474	0.11111	0.59969
4	0.44048	0.55952		− 0.1491	− 0.5807	− 0.5436	0.19290	0.71076
5	0.55952	0.44048		0.1494	− 0.8199	− 0.1986	0.28924	0.80710
6	0.67857	0.32143		0.4633	− 1.1350	0.1266	0.40031	0.88889
7	0.79762	0.20238		0.8330	− 1.5976	0.4685	0.52932	0.95361
8	0.91667	0.08333		1.3832	− 2.4849	0.9102	0.68766	0.99361
			S3:	0.0671	− 7.3189	− 4.1125		
			S4:	5.5253	11.2338	10.4686		
Total number of results: 9								
1	0.07447	0.92553		− 1.3849	− 0.0774	− 2.5589	0.00568	0.28313
2	0.18085	0.81915		− 0.8964	− 0.1995	− 1.6120	0.04102	0.42914
3	0.28723	0.71277		− 0.5563	− 0.3386	− 1.0829	0.09775	0.54964
4	0.39362	0.60638		− 0.2685	− 0.5002	− 0.6927	0.16875	0.65506
5	0.50000	0.50000		0.0000	− 0.6931	− 0.3665	0.25137	0.74863
6	0.60638	0.39362		0.2695	− 0.9324	− 0.0700	0.34494	0.83125
7	0.71277	0.28723		0.5611	− 1.2475	0.2211	0.45036	0.90225
8	0.81915	0.18085		0.9120	− 1.7101	0.5365	0.57086	0.95898
9	0.92553	0.07447		1.4436	− 2.5974	0.9545	0.71687	0.99432
			S3:	0.0813	− 8.2961	− 4.6709		
			S4:	6.4606	12.9874	12.1863		
Total number of results: 10								
1	0.06731	0.93269		− 1.4307	− 0.0697	− 2.6638	0.00512	0.25887
2	0.16346	0.83654		− 0.9613	− 0.1785	− 1.7233	0.03677	0.39416
3	0.25962	0.74038		− 0.6376	− 0.3006	− 1.2020	0.08726	0.50690
4	0.35577	0.64423		− 0.3674	− 0.4397	− 0.8217	0.15003	0.60662
5	0.45192	0.54808		− 0.1203	− 0.6013	− 0.5086	0.22244	0.69646
6	0.54808	0.45192		0.1205	− 0.7942	− 0.2304	0.30354	0.77756
7	0.64423	0.35577		0.3693	− 1.0335	0.0329	0.39338	0.84997
8	0.74038	0.25962		0.6442	− 1.3486	0.2990	0.49310	0.91274
9	0.83654	0.16346		0.9803	− 1.8112	0.5940	0.60584	0.96323
10	0.93269	0.06731		1.4964	− 2.6985	0.9927	0.74113	0.99488
			S3:	0.0934	− 9.2757	− 5.2311		
			S4:	7.2933	14.7617	13.9262		
Total number of results: 11								
1	0.06140	0.93860		− 1.4709	− 0.0634	− 2.7588	0.00465	0.23840
2	0.14912	0.85088		− 1.0179	− 0.1615	− 1.8233	0.03332	0.36436
3	0.23684	0.76316		− 0.7077	− 0.2703	− 1.3083	0.07882	0.47009
4	0.32456	0.67544		− 0.4515	− 0.3924	− 0.9355	0.13507	0.56437
5	0.41228	0.58772		− 0.2206	− 0.5315	− 0.6320	0.19958	0.65019
6	0.50000	0.50000		0.0000	− 0.6931	− 0.3665	0.27125	0.72875
7	0.58772	0.41228		0.2213	− 0.8861	− 0.1210	0.34981	0.80042

Table T2 *Cont.*

i	F'(i)	1 − F'(i)	Xn(i)	Xe(i)	Xw(i)	p0.05	p0.95
8	0.67544	0.32456	0.4546	−1.1253	0.1180	0.43563	0.86492
9	0.76316	0.23684	0.7162	−1.4404	0.3649	0.52991	0.92118
10	0.85088	0.14912	1.0402	−1.9030	0.6434	0.63564	0.96668
11	0.93860	0.06140	1.5434	−2.7903	1.0261	0.76160	0.99535
S3:			0.1070	−10.2572	−5.7928		
S4:			8.1856	16.5531	15.6845		

Total number of results: 12

i	F'(i)	1 − F'(i)	Xn(i)	Xe(i)	Xw(i)	p0.05	p0.95
1	0.05645	0.94355	−1.5065	−0.0581	−2.8455	0.00426	0.22092
2	0.13710	0.86290	−1.0678	−0.1475	−1.9142	0.03046	0.33868
3	0.21774	0.78226	−0.7691	−0.2456	−1.4042	0.07187	0.43811
4	0.29839	0.70161	−0.5244	−0.3544	−1.0374	0.12285	0.52733
5	0.37903	0.62097	−0.3063	−0.4765	−0.7413	0.18102	0.60914
6	0.45968	0.54032	−0.1009	−0.6156	−0.4852	0.24530	0.68476
7	0.54032	0.45968	0.1010	−0.7772	−0.2520	0.31524	0.75470
8	0.62097	0.37903	0.3076	−0.9701	−0.0303	0.39086	0.81898
9	0.70161	0.29839	0.5286	−1.2094	0.1901	0.47267	0.87715
10	0.78226	0.21774	0.7796	−1.5244	0.4216	0.56189	0.92813
11	0.86290	0.13710	1.0935	−1.9871	0.6867	0.66132	0.96954
12	0.94355	0.05645	1.5856	−2.8744	1.0558	0.77908	0.99573
S3:			0.1209	−11.2402	−6.3559		
S4:			9.0822	18.3592	17.4585		

Total number of results: 13

i	F'(i)	1 − F'(i)	Xn(i)	Xe(i)	Xw(i)	p0.05	p0.95
1	0.05224	0.94776	−1.5383	−0.0537	−2.9252	0.00394	0.20582
2	0.12687	0.87313	−1.1124	−0.1357	−1.9976	0.02805	0.31634
3	0.20149	0.79851	−0.8236	−0.2250	−1.4916	0.07187	0.41010
4	0.27612	0.72388	−0.5886	−0.3231	−1.1297	0.11267	0.49465
5	0.35075	0.64925	−0.3808	−0.4319	−0.8395	0.16566	0.57262
6	0.42537	0.57463	−0.1873	−0.5540	−0.5905	0.22395	0.64520
7	0.50000	0.50000	0.0000	−0.6931	−0.3665	0.28705	0.71295
8	0.57463	0.42537	0.1878	−0.8548	−0.1569	0.35480	0.77604
9	0.64925	0.35075	0.3829	−1.0477	0.0466	0.42738	0.83434
10	0.72388	0.27612	0.5941	−1.2869	0.2523	0.50535	0.88733
11	0.79851	0.20149	0.8361	−1.6020	0.4713	0.58990	0.93395
12	0.87313	0.12687	1.1414	−2.0646	0.7249	0.68366	0.97195
13	0.94776	0.05224	1.6239	−2.9519	1.0825	0.79418	0.99606
S3:			0.1351	−12.2245	−6.9200		
S4:			9.9824	20.1778	19.2459		

Total number of results: 14

i	F'(i)	1 − F'(i)	Xn(i)	Xe(i)	Xw(i)	p0.05	p0.95
1	0.04861	0.95139	−1.5671	−0.0498	−2.9991	0.00366	0.19264
2	0.11806	0.88194	−1.1526	−0.1256	−2.0744	0.02600	0.29673
3	0.18750	0.81250	−0.8724	−0.2076	−1.5720	0.06110	0.38539
4	0.25694	0.74306	−0.6457	−0.2970	−1.2141	0.10405	0.46566
5	0.32639	0.67361	−0.4465	−0.3951	−0.9286	0.15272	0.54000
6	0.39583	0.60417	−0.2628	−0.5039	−0.6854	0.20607	0.60928
7	0.46528	0.53472	−0.0868	−0.6260	−0.4684	0.26358	0.67497
8	0.53472	0.46528	0.0869	−0.7651	−0.2677	0.32503	0.73641
9	0.60417	0.39583	0.2637	−0.9268	−0.0761	0.39041	0.79393
10	0.67361	0.32639	0.4495	−1.1197	0.1130	0.45999	0.84728

Table T2 *Cont.*

i	$F'(i)$	$1 - F'(i)$		Xn(i)	Xe(i)	Xw(i)	p0.05	p0.95
11	0.74306	0.25694		0.6525	−1.3589	0.3067	0.53434	0.89595
12	0.81250	0.18750		0.8870	−1.6740	0.5152	0.61461	0.93890
13	0.88194	0.11806		1.1849	−2.1366	0.7592	0.70327	0.97400
14	0.95139	0.04861		1.6588	−3.0239	1.1065	0.80736	0.99634
			S3:	0.1494	−13.2100	−7.4850		
			S4:	10.8856	22.0073	21.0449		

Total number of results: 15

i	$F'(i)$	$1 - F'(i)$		Xn(i)	Xe(i)	Xw(i)	p0.05	p0.95
1	0.04545	0.95455		−1.5932	−0.0465	−3.0679	0.00341	0.18104
2	0.11039	0.88961		−1.1891	−0.1170	−2.1458	0.02423	0.27940
3	0.17532	0.82468		−0.9166	−0.1928	−1.6463	0.05685	0.36344
4	0.24026	0.75974		−0.6970	−0.2748	−1.2918	0.09666	0.43978
5	0.30519	0.69481		−0.5052	−0.3641	−1.0103	0.14166	0.51075
6	0.37013	0.62987		−0.3295	−0.4622	−0.7717	0.19086	0.51075
7	0.43506	0.56494		−0.1628	−0.5710	−0.5603	0.24373	0.57744
8	0.50000	0.50000		0.0000	−0.6931	−0.3665	0.29999	0.64043
9	0.56494	0.43506		0.1632	−0.8323	−0.1836	0.35956	0.70001
10	0.62987	0.37013		0.3311	−0.9939	−0.0061	0.42256	0.75627
11	0.69481	0.30519		0.5091	−1.1868	0.1713	0.48925	0.80913
12	0.75974	0.24026		0.7052	−1.4260	0.3549	0.56022	0.85834
13	0.82468	0.17532		0.9332	−1.7411	0.5545	0.63656	0.90334
14	0.88961	0.11039		1.2246	−2.2037	0.7902	0.72060	0.94315
15	0.95455	0.04545		1.6910	−3.0910	1.1285	0.81896	0.99659
			S3:	0.1639	−14.1965	−8.0587		
			S4:	11.7913	23.8464	22.8541		

Total number of results: 16

i	$F'(i)$	$1 - F'(i)$		Xn(i)	Xe(i)	Xw(i)	p0.05	p0.95
1	0.04268	0.95732		−1.6172	−0.0436	−3.1322	0.00320	0.17075
2	0.10366	0.89634		−1.2224	−0.1094	−2.2124	0.02268	0.26396
3	0.16463	0.83537		−0.9568	−0.1799	−1.7154	0.05315	0.34383
4	0.22561	0.77439		−0.7435	−0.2557	−1.3638	0.09025	0.41657
5	0.28659	0.71341		−0.5582	−0.3377	−1.0856	0.13211	0.48440
6	0.34756	0.65244		−0.3893	−0.4270	−0.8509	0.17777	0.54835
7	0.40854	0.59146		−0.2302	−0.5252	−0.6441	0.22669	0.60899
8	0.46951	0.53049		−0.0762	−0.6340	−0.4558	0.27860	0.66663
9	0.53049	0.46951		0.0763	−0.7561	−0.2796	0.33337	0.72140
10	0.59146	0.40854		0.2309	−0.8952	−0.1107	0.39101	0.77331
11	0.65244	0.34756		0.3915	−1.0568	0.0553	0.45165	0.82223
12	0.71341	0.28659		0.5630	−1.2497	0.2229	0.51560	0.86789
13	0.77439	0.22561		0.7531	−1.4889	0.3981	0.58343	0.90975
14	0.83537	0.16463		0.9755	−1.8040	0.5900	0.65617	0.94685
15	0.89634	0.10366		1.2611	−2.2667	0.8183	0.73604	0.97732
16	0.95732	0.04268		1.7207	−3.1540	1.1487	0.82925	0.99680
			S3:	0.1785	−15.1838	−8.6174		
			S4:	12.6993	25.6940	24.6724		

Total number of results: 17

i	$F'(i)$	$1 - F'(i)$		Xn(i)	Xe(i)	Xw(i)	p0.05	p0.95
1	0.04023	0.95977		−1.6392	−0.0411	−3.1927	0.00301	0.16157
2	0.09770	0.90230		−1.2530	−0.1028	−2.2749	0.02132	0.25012
3	0.15517	0.84483		−0.9936	−0.1686	−1.7801	0.04990	0.32619
4	0.21264	0.78736		−0.7860	−0.2391	−1.4310	0.08464	0.39564

Table T2 *Cont.*

i	F'(i)	1 − F'(i)		Xn(i)	Xe(i)	Xw(i)	p0.05	p0.95
5	0.27011	0.72989		− 0.6062	− 0.3149	− 1.1556	0.12377	0.46055
6	0.32759	0.67241		− 0.4433	− 0.3969	− 0.9241	0.16636	0.52192
7	0.38506	0.61494		− 0.2906	− 0.4862	− 0.7211	0.21191	0.58029
8	0.44253	0.55747		− 0.1440	− 0.5843	− 0.5373	0.26011	0.63599
9	0.50000	0.50000		0.0000	− 0.6931	− 0.3665	0.31083	0.68917
10	0.55747	0.44253		0.1443	− 0.8152	− 0.2043	0.36401	0.73989
11	0.61494	0.38506		0.2918	− 0.9544	− 0.0467	0.41970	0.78809
12	0.67241	0.32759		0.4462	− 1.1160	0.1098	0.47808	0.83364
13	0.72989	0.27011		0.6121	− 1.3089	0.2692	0.53945	0.87623
14	0.78736	0.21264		0.7971	− 1.5481	0.4371	0.60436	0.91535
15	0.84483	0.15517		1.0145	− 1.8632	0.6223	0.67381	0.95010
16	0.90230	0.09770		1.2949	− 2.3258	0.8441	0.74988	0.97869
17	0.95977	0.04023		1.7484	− 3.2131	1.1673	0.83843	0.99699
			S3:	0.1933	− 16.1719	− 9.1845		
			S4:	13.6092	27.5491	26.4987		

Total number of results: 18

i	F'(i)	1 − F'(i)		Xn(i)	Xe(i)	Xw(i)	p0.05	p0.95
1	0.03804	0.96196		− 1.6595	− 0.0388	− 3.2497	0.00285	0.15332
2	0.09239	0.90761		− 1.2812	− 0.0969	− 2.3336	0.02011	0.23766
3	0.14674	0.85326		− 1.0276	− 0.1587	− 1.8408	0.04702	0.31026
4	0.20109	0.79891		− 0.8250	− 0.2245	− 1.4939	0.07696	0.37668
5	0.25543	0.74457		− 0.6502	− 0.2950	− 1.2209	0.11643	0.43888
6	0.30978	0.69022		− 0.4924	− 0.3707	− 0.9922	0.15634	0.49783
7	0.36413	0.63587		− 0.3453	− 0.4528	− 0.7924	0.19895	0.55404
8	0.41848	0.58152		− 0.2048	− 0.5421	− 0.6123	0.24396	0.69784
9	0.47283	0.52717		− 0.0679	− 0.6402	− 0.4459	0.29120	0.65940
10	0.52717	0.47283		0.0680	− 0.7490	− 0.2890	0.34060	0.70880
11	0.58152	0.41848		0.2054	− 0.8711	− 0.1380	0.39215	0.75604
12	0.63587	0.36413		0.3470	− 1.0102	0.0102	0.44595	0.80105
13	0.69022	0.30978		0.4961	− 1.1719	0.1586	0.50217	0.84366
14	0.74457	0.25543		0.6572	− 1.3648	0.3110	0.56112	0.88357
15	0.79891	0.20109		0.8376	− 1.6040	0.4725	0.62332	0.92030
16	0.85326	0.14674		1.0505	− 1.9191	0.6519	0.68974	0.95297
17	0.90761	0.09239		1.3264	− 2.3817	0.8678	0.76234	0.97989
18	0.96196	0.03804		1.7743	− 3.2690	1.1845	0.84668	0.99715
			S3:	0.2083	− 17.1607	− 9.7522		
			S4:	14.5208	29.4111	28.3323		

Total number of results: 19

i	F'(i)	1 − F'(i)		Xn(i)	Xe(i)	Xw(i)	p0.05	p0.95
1	0.03608	0.96392		− 1.6784	− 0.0367	− 3.3036	0.00270	0.14587
2	0.08763	0.91237		− 1.3075	− 0.0917	− 2.3891	0.01903	0.22637
3	0.13918	0.86082		− 1.0590	− 0.1499	− 1.8980	0.04446	0.29580
4	0.19072	0.80928		− 0.8610	− 0.2116	− 1.5530	0.07529	0.35943
5	0.24227	0.75773		− 0.6907	− 0.2774	− 1.2822	0.10991	0.41912
6	0.29381	0.70619		− 0.5374	− 0.3479	− 1.0559	0.14747	0.47580
7	0.34536	0.65464		− 0.3952	− 0.4237	− 0.8588	0.18750	0.52997
8	0.39691	0.60309		− 0.2600	− 0.5057	− 0.6818	0.22972	0.58194
9	0.44845	0.55155		− 0.1291	− 0.5950	− 0.5191	0.27395	0.63188
10	0.50000	0.50000		0.0000	− 0.6931	− 0.3665	0.32009	0.67991
11	0.55155	0.44845		0.1293	− 0.8019	− 0.2207	0.36811	0.72605

Table T2 *Cont.*

i	F'(i)	1 − F'(i)		Xn(i)	Xe(i)	Xw(i)	p0.05	p0.95
12	0.60309	0.39691		0.2609	−0.9241	−0.0790	0.41806	0.77028
13	0.65464	0.34536		0.3974	−1.0632	0.0613	0.47003	0.81250
14	0.70619	0.29381		0.5419	−1.2248	0.2028	0.52420	0.85253
15	0.75773	0.24227		0.6987	−1.4177	0.3490	0.58088	0.89009
16	0.80928	0.19072		0.8751	−1.6569	0.5050	0.64057	0.92471
17	0.86082	0.13918		1.0841	−1.9720	0.6791	0.70420	0.95553
18	0.91237	0.08763		1.3557	−2.4346	0.8898	0.77363	0.98097
19	0.96392	0.03608		1.7985	−3.3219	1.2006	0.85413	0.99730
			S3:	0.2233	−18.1500	−10.3204		
			S4:	15.4339	31.2792	30.1724		

Total number of results: 20

i	F'(i)	1 − F'(i)		Xn(i)	Xe(i)	Xw(i)	p0.05	p0.95
1	0.03431	0.96569		−1.6960	−0.0349	−3.3548	0.00256	0.13911
2	0.08333	0.91667		−1.3319	−0.0870	−2.4417	0.01806	0.21611
3	0.13235	0.86765		−1.0883	−0.1420	−1.9521	0.04217	0.28262
4	0.18137	0.81863		−0.8945	−0.2001	−1.6088	0.07135	0.34366
5	0.23039	0.76961		−0.7282	−0.2619	−1.3399	0.10408	0.40103
6	0.27941	0.72059		−0.5789	−0.3277	−1.1157	0.13955	0.45558
7	0.32843	0.67157		−0.4410	−0.3981	−0.9210	0.17731	0.50782
8	0.37745	0.62255		−0.3104	−0.4739	−0.7467	0.21707	0.55803
9	0.42647	0.57353		−0.1845	−0.5559	−0.5871	0.25865	0.60641
10	0.47549	0.52451		−0.0613	−0.6453	−0.4381	0.30195	0.65307
11	0.52451	0.47549		0.0613	−0.7434	−0.2965	0.34693	0.69805
12	0.57353	0.42647		0.1850	−0.8522	−0.1599	0.39358	0.74135
13	0.62255	0.37745		0.3117	−0.9743	−0.0260	0.44197	0.78293
14	0.67157	0.32843		0.4438	−1.1134	0.1074	0.49218	0.82269
15	0.72059	0.27941		0.5842	−1.2751	0.2430	0.54442	0.86045
16	0.76961	0.23039		0.7373	−1.4680	0.3839	0.59897	0.89592
17	0.81863	0.18137		0.9100	−1.7072	0.5349	0.65634	0.92865
18	0.86765	0.13235		1.1154	−2.0223	0.7042	0.71738	0.95783
19	0.91667	0.08333		1.3832	−2.4849	0.9102	0.78389	0.98193
20	0.96569	0.03431		1.8213	−3.3722	1.2156	0.86089	0.99744
			S3:	0.2384	−19.1399	−10.8891		
			S4:	16.3484	33.1529	32.0184		

Total number of results: 21

i	F'(i)	1 − F'(i)		Xn(i)	Xe(i)	Xw(i)	p0.05	p0.95
1	0.03271	0.96729		−1.7124	−0.0333	−3.4035	0.00244	0.13295
2	0.07944	0.92056		−1.3547	−0.0828	−2.4917	0.01719	0.20673
3	0.12617	0.87383		−1.1156	−0.1349	−2.0035	0.04010	0.27055
4	0.17290	0.82710		−0.9256	−0.1898	−1.6616	0.06781	0.32921
5	0.21963	0.78037		−0.7629	−0.2480	−1.3944	0.09884	0.38441
6	0.26636	0.73364		−0.6174	−0.3097	−1.1721	0.13245	0.43698
7	0.31308	0.68692		−0.4832	−0.3755	−0.9794	0.16818	0.48739
8	0.35981	0.64019		−0.3567	−0.4460	−0.8074	0.20575	0.53594
9	0.40654	0.59346		−0.2353	−0.5218	−0.6505	0.24499	0.58280
10	0.45327	0.54673		−0.1170	−0.6038	−0.5045	0.28580	0.62810
11	0.50000	0.50000		0.0000	−0.6931	−0.3665	0.32811	0.67189
12	0.54673	0.45327		0.1171	−0.7913	−0.2341	0.37190	0.71420
13	0.59346	0.40654		0.2360	−0.9001	−0.1053	0.41720	0.75501
14	0.64019	0.35981		0.3585	−1.0222	0.0219	0.46406	0.79425

Table T2 *Cont.*

i	F'(i)	1 − F'(i)		Xn(i)	Xe(i)	Xw(i)	p0.05	p0.95
15	0.68692	0.31308		0.4867	−1.1613	0.1495	0.51261	0.83182
16	0.73364	0.26636		0.6235	−1.3229	0.2798	0.56302	0.86755
17	0.78037	0.21963		0.7732	−1.5158	0.4160	0.61559	0.90116
18	0.82710	0.17290		0.9427	−1.7551	0.5625	0.67079	0.93219
19	0.87383	0.12617		1.1448	−2.0701	0.7276	0.72945	0.95990
20	0.92056	0.07944		1.4091	−2.5328	0.9293	0.79327	0.98281
21	0.96729	0.03271		1.8428	−3.4201	1.2297	0.86705	0.99756
			S3:	0.2537	−20.1303	−11.4581		
			S4:	17.2641	35.0318	33.8698		

Total number of results: 22

i	F'(i)	1 − F'(i)		Xn(i)	Xe(i)	Xw(i)	p0.05	p0.95
1	0.03125	0.96875		−1.7279	−0.0317	−3.4499	0.00233	0.12731
2	0.07589	0.92411		−1.3761	−0.0789	−2.5392	0.01640	0.19812
3	0.12054	0.87946		−1.1411	−0.1284	−2.0523	0.03822	0.25947
4	0.16518	0.83482		−0.9547	−0.1805	−1.7118	0.06460	0.31591
5	0.20982	0.79018		−0.7954	−0.2355	−1.4461	0.09411	0.36909
6	0.25446	0.74554		−0.6532	−0.2937	−1.2254	0.12603	0.41980
7	0.29911	0.70089		−0.5224	−0.3554	−1.0345	0.15994	0.46849
8	0.34375	0.65625		−0.3995	−0.4212	−0.8646	0.19556	0.51546
9	0.38839	0.61161		−0.2820	−0.4917	−0.7100	0.23272	0.56087
10	0.43304	0.56696		−0.1679	−0.5675	−0.5666	0.27131	0.60484
11	0.47768	0.52232		−0.0558	−0.6495	−0.4316	0.31126	0.64746
12	0.52232	0.47768		0.0558	−0.7388	−0.3027	0.35254	0.68874
13	0.56696	0.43304		0.1683	−0.8369	−0.1780	0.39516	0.72869
14	0.61161	0.38839		0.2831	−0.9457	−0.0558	0.43913	0.76728
15	0.65625	0.34375		0.4018	−1.0678	0.0656	0.48454	0.80444
16	0.70089	0.29911		0.5266	−1.2070	0.1881	0.53151	0.84006
17	0.74554	0.25446		0.6602	−1.3686	0.3138	0.58020	0.87397
18	0.79018	0.20982		0.8068	−1.5615	0.4456	0.63091	0.90589
19	0.83482	0.16518		0.9733	−1.8007	0.5882	0.68409	0.93540
20	0.87946	0.12054		1.1724	−2.1158	0.7494	0.74053	0.96178
21	0.92411	0.07589		1.4335	−2.5784	0.9472	0.80188	0.98360
22	0.96875	0.03125		1.8631	−3.4657	1.2429	0.87269	0.99767
			S3:	0.2690	−21.1211	−12.0275		
			S4:	18.1809	36.9153	35.7262		

Total number of results: 23

i	F'(i)	1 − F'(i)		Xn(i)	Xe(i)	Xw(i)	p0.05	p0.95
1	0.02991	0.97009		−1.7424	−0.0304	−3.4943	0.00223	0.12212
2	0.07265	0.92735		−1.3963	−0.0754	−2.5846	0.01567	0.19020
3	0.11538	0.88462		−1.1651	−0.1226	−2.0988	0.03651	0.24925
4	0.15812	0.84188		−0.9820	−0.1721	−1.7596	0.06167	0.30364
5	0.20085	0.79915		−0.8258	−0.2242	−1.4952	0.08981	0.35193
6	0.24359	0.75641		−0.6866	−0.2792	−1.2759	0.12021	0.40390
7	0.28632	0.71368		−0.5589	−0.3373	−1.0867	0.15248	0.45097
8	0.32906	0.67094		−0.4392	−0.3991	−0.9186	0.18634	0.49643
9	0.37179	0.62821		−0.3252	−0.4649	−0.7660	0.22164	0.54046
10	0.41453	0.58547		−0.2149	−0.5353	−0.6249	0.25824	0.58315
11	0.45727	0.54274		−0.1069	−0.6111	−0.4924	0.29609	0.62461
12	0.50000	0.50000		0.0000	−0.6931	−0.3665	0.33515	0.66485
13	0.54274	0.45727		0.1071	−0.7825	−0.2453	0.37539	0.70391

Table T2 *Cont.*

i	F'(i)	1 − F'(i)		Xn(i)	Xe(i)	Xw(i)	p0.05	p0.95
14	0.58547	0.41453		0.2155	−0.8806	−0.1271	0.41684	0.74176
15	0.62821	0.37179		0.3267	−0.9894	−0.0106	0.45954	0.77836
16	0.67094	0.32906		0.4421	−1.1115	0.1057	0.50356	0.81366
17	0.71368	0.28632		0.5638	−1.2506	0.2236	0.54902	0.84752
18	0.75641	0.24359		0.6945	−1.4123	0.3452	0.59610	0.87978
19	0.79915	0.20085		0.8384	−1.6052	0.4732	0.64507	0.91019
20	0.84188	0.15812		1.0022	−1.8444	0.6122	0.69636	0.93832
21	0.88462	0.11538		1.1985	−2.1595	0.7699	0.75075	0.96348
22	0.92735	0.07265		1.4566	−2.6221	0.9640	0.80980	0.98433
23	0.97009	0.02991		1.8825	−3.5094	1.2554	0.87788	0.99777
			S3:	0.2844	−22.1123	−12.5973		
			S4:	19.0987	38.8031	37.5871		

Total number of results: 24

i	F'(i)	1 − F'(i)		Xn(i)	Xe(i)	Xw(i)	p0.05	p0.95
1	0.02869	0.97131		−1.7560	−0.0291	−3.5367	0.00213	0.11735
2	0.06967	0.93033		−1.4153	−0.0722	−2.6281	0.01501	0.18289
3	0.11066	0.88934		−1.1878	−0.1173	−2.1433	0.03495	0.23980
4	0.15164	0.84836		−1.0077	−0.1644	−1.8052	0.05901	0.29227
5	0.19262	0.80738		−0.8543	−0.2140	−1.5419	0.08588	0.34181
6	0.23361	0.76639		−0.7179	−0.2661	−1.3240	0.11491	0.38914
7	0.27459	0.72541		−0.5931	−0.3210	−1.1363	0.14569	0.43469
8	0.31557	0.68443		−0.4763	−0.3792	−0.9698	0.17796	0.47853
9	0.35656	0.64344		−0.3653	−0.4409	−0.8189	0.21157	0.52142
10	0.39754	0.60246		−0.2584	−0.5067	−0.6798	0.24639	0.56289
11	0.43852	0.56148		−0.1541	−0.5772	−0.5496	0.28236	0.60321
12	0.47951	0.52049		−0.0512	−0.6530	−0.4262	0.31942	0.64244
13	0.52049	0.47951		0.0513	−0.7350	−0.3079	0.35756	0.68058
14	0.56148	0.43852		0.1544	−0.8243	−0.1932	0.39678	0.71764
15	0.60246	0.39754		0.2593	−0.9225	−0.0807	0.43711	0.75361
16	0.64344	0.35656		0.3672	−1.0313	0.0308	0.47858	0.78843
17	0.68443	0.31557		0.4797	−1.1534	0.1427	0.52127	0.82204
18	0.72541	0.27459		0.5986	−1.2925	0.2566	0.56531	0.85431
19	0.76639	0.23361		0.7268	−1.4541	0.3744	0.61086	0.88509
20	0.80738	0.19262		0.8681	−1.6470	0.4990	0.65819	0.91411
21	0.84836	0.15164		1.0294	−1.8863	0.6346	0.70773	0.94099
22	0.88934	0.11066		1.2232	−2.2013	0.7891	0.76020	0.96505
23	0.93033	0.06967		1.4785	−2.6640	0.9798	0.81711	0.98499
24	0.97131	0.02869		1.9008	−3.5513	1.2673	0.88265	0.99786
			S3:	0.2999	−23.1039	−13.1673		
			S4:	20.0174	40.6949	39.4522		

Total number of results: 25

i	F'(i)	1 − F'(i)		Xn(i)	Xe(i)	Xw(i)	p0.05	p0.95
1	0.02756	0.97244		−1.7690	−0.0279	−3.5775	0.00205	0.11293
2	0.06693	0.93307		−1.4332	−0.0693	−2.6697	0.01440	0.17612
3	0.10630	0.89370		−1.2091	−0.1124	−2.1858	0.03352	0.23104
4	0.14567	0.85433		−1.0320	−0.1574	−1.8487	0.05656	0.28172
5	0.18504	0.81496		−0.8812	−0.2046	−1.5866	0.08229	0.32961
6	0.22441	0.77559		−0.7474	−0.2541	−1.3699	0.11006	0.37541
7	0.26378	0.73622		−0.6251	−0.3062	−1.1834	0.13947	0.41952
8	0.30315	0.69685		−0.5110	−0.3612	−1.0184	0.17030	0.46221

Table T2 *Cont.*

i	F'(i)	1 − F'(i)	Xn(i)	Xe(i)	Xw(i)	p0.05	p0.95
9	0.34252	0.65748	−0.4028	−0.4193	−0.8691	0.20238	0.50364
10	0.38189	0.61811	−0.2988	−0.4811	−0.7317	0.23559	0.54393
11	0.42126	0.57874	−0.1978	−0.5469	−0.6035	0.26985	0.58316
12	0.46063	0.53937	−0.0985	−0.6174	−0.4823	0.30513	0.62138
13	0.50000	0.50000	0.0000	−0.6931	−0.3665	0.34139	0.65861
14	0.53937	0.46063	0.0986	−0.7752	−0.2547	0.37862	0.69487
15	0.57874	0.42126	0.1983	−0.8645	−0.1456	0.41684	0.73015
16	0.61811	0.38189	0.3001	−0.9626	−0.0381	0.45607	0.76441
17	0.65748	0.34252	0.4052	−1.0714	0.0690	0.49636	0.79762
18	0.69685	0.30315	0.5150	−1.1935	0.1769	0.53779	0.82970
19	0.73622	0.26378	0.6314	−1.3326	0.2872	0.58048	0.86052
20	0.77559	0.22441	0.7572	−1.4943	0.4016	0.62459	0.88994
21	0.81496	0.18504	0.8962	−1.6872	0.5231	0.67039	0.91771
22	0.85433	0.14567	1.0552	−1.9264	0.6557	0.71828	0.94344
23	0.89370	0.10630	1.2466	−2.2415	0.8071	0.76896	0.96648
24	0.93307	0.06693	1.4994	−2.7041	0.9948	0.82388	0.98560
25	0.97244	0.02756	1.9184	−3.5914	1.2785	0.88707	0.99795
		S3:	0.3155	−24.0958	−13.7376		
		S4:	20.9370	42.5904	41.3212		

Total number of results: 26

i	F'(i)	1 − F'(i)	Xn(i)	Xe(i)	Xw(i)	p0.05	p0.95
1	0.02652	0.97348	−1.7812	−0.0269	−3.6166	0.00197	0.10883
2	0.06439	0.93561	−1.4502	−0.0666	−2.7096	0.01384	0.16983
3	0.10227	0.89773	−1.2294	−0.1079	−2.2267	0.03220	0.22289
4	0.14015	0.85985	−1.0549	−0.1510	−1.8905	0.05431	0.27190
5	0.17803	0.82197	−0.9066	−0.1961	−1.6294	0.07899	0.31824
6	0.21591	0.78409	−0.7752	−0.2432	−1.4137	0.10560	0.36260
7	0.25379	0.74621	−0.6552	−0.2927	−1.2285	0.13377	0.40535
8	0.29167	0.70833	−0.5436	−0.3448	−1.0647	0.16328	0.44677
9	0.32955	0.67045	−0.4379	−0.3998	−0.9168	0.19396	0.48700
10	0.36742	0.63258	−0.3366	−0.4580	−0.7810	0.22570	0.52616
11	0.40530	0.59470	−0.2385	−0.5197	−0.6545	0.25842	0.56434
12	0.44318	0.55682	−0.1423	−0.5855	−0.5353	0.29508	0.60158
13	0.48106	0.51894	−0.0473	−0.6560	−0.4216	0.32664	0.63791
14	0.51894	0.48106	0.0474	−0.7318	−0.3123	0.36209	0.67336
15	0.55682	0.44318	0.1426	−0.8138	−0.2061	0.39842	0.70792
16	0.59470	0.40530	0.2392	−0.9031	−0.1019	0.43566	0.74158
17	0.63258	0.36742	0.3382	−1.0012	0.0012	0.47384	0.77430
18	0.67045	0.32955	0.4407	−1.1100	0.1044	0.51300	0.80604
19	0.70833	0.29167	0.5481	−1.2321	0.2088	0.55323	0.83672
20	0.74621	0.25379	0.6623	−1.3713	0.3157	0.59465	0.86623
21	0.78409	0.21591	0.7859	−1.5329	0.4272	0.63740	0.89440
22	0.82197	0.17803	0.9228	−1.7258	0.5457	0.68176	0.92101
23	0.85985	0.14015	1.0797	−1.9650	0.6755	0.72810	0.94569
24	0.89773	0.10227	1.2689	−2.2801	0.8242	0.77711	0.96780
25	0.93561	0.06439	1.5192	−2.7427	1.0090	0.83017	0.98616
26	0.97348	0.02652	1.9351	−3.6300	1.2892	0.89117	0.99803
		S3:	0.3311	−25.0881	−14.3082		
		S4:	21.8573	44.4893	43.1939		

Table T2 *Cont.*

i	F'(i)	1 − F'(i)	Xn(i)	Xe(i)	Xw(i)	p0.05	p0.95
Total number of results: 27							
1	0.02555	0.97445	−1.7928	−0.0259	−3.6543	0.00190	0.10502
2	0.06204	0.93796	−1.4664	−0.0641	−2.7481	0.01332	0.16397
3	0.09854	0.90146	−1.2486	−0.1037	−2.2659	0.03098	0.21530
4	0.13504	0.86496	−1.0767	−0.1451	−1.9306	0.05223	0.26274
5	0.17153	0.82847	−0.9307	−0.1882	−1.6704	0.07594	0.30763
6	0.20803	0.79197	−0.8014	−0.2332	−1.4557	0.10148	0.35062
7	0.24453	0.75547	−0.6837	−0.2804	−1.2715	0.12852	0.39210
8	0.28102	0.71898	−0.5743	−0.3299	−1.1089	0.15682	0.43230
9	0.31752	0.68248	−0.4709	−0.3820	−0.9623	0.18622	0.47139
10	0.35401	0.64599	−0.3721	−0.4370	−0.8279	0.21662	0.50489
11	0.39051	0.60949	−0.2765	−0.4951	−0.7029	0.24793	0.54664
12	0.42701	0.57299	−0.1832	−0.5569	−0.5854	0.28012	0.58293
13	0.46350	0.53650	−0.0913	−0.6227	−0.4737	0.31314	0.61839
14	0.50000	0.50000	0.0000	−0.6931	−0.3665	0.34697	0.65303
15	0.53650	0.46350	0.0914	−0.7689	−0.2627	0.38161	0.68686
16	0.57299	0.42701	0.1836	−0.8510	−0.1614	0.41707	0.71988
17	0.60949	0.39051	0.2776	−0.9403	−0.0616	0.45336	0.75207
18	0.64599	0.35401	0.3741	−1.0384	0.0377	0.49052	0.78338
19	0.68248	0.31752	0.4742	−1.1472	0.1373	0.52861	0.81378
20	0.71898	0.28102	0.5794	−1.2693	0.2385	0.56770	0.84318
21	0.75547	0.24453	0.6915	−1.4084	0.3425	0.60790	0.87148
22	0.79197	0.20803	0.8131	−1.5701	0.4511	0.64936	0.89851
23	0.82847	0.17153	0.9481	−1.7630	0.5670	0.69237	0.92406
24	0.86496	0.13504	1.1029	−2.0022	0.6943	0.73726	0.94777
25	0.90146	0.09854	1.2901	−2.3173	0.8404	0.78470	0.96902
26	0.93796	0.06204	1.5382	−2.7799	1.0224	0.83603	0.98668
27	0.97445	0.02555	1.9511	−3.6672	1.2994	0.89498	0.99810
		S3:	0.3467	−26.0806	−14.8790		
		S4:	22.7783	46.3914	45.0699		
Total number of results: 28							
1	0.02465	0.97535	−1.8039	−0.0250	−3.6906	0.00183	0.10147
2	0.05986	0.94014	−1.4818	−0.0617	−2.7851	0.01284	0.15851
3	0.09507	0.90493	−1.2668	−0.0999	−2.3036	0.02985	0.20821
4	0.13028	0.86972	−1.0973	−0.1396	−1.9691	0.05031	0.25417
5	0.16549	0.83451	−0.9535	−0.1809	−1.7097	0.07311	0.29769
6	0.20070	0.79930	−0.8263	−0.2240	−1.4960	0.09768	0.33940
7	0.23592	0.76408	−0.7106	−0.2691	−1.3128	0.12367	0.37967
8	0.27113	0.72887	−0.6033	−0.3163	−1.1512	0.15085	0.41873
9	0.30634	0.69366	−0.5020	−0.3658	−1.0057	0.17908	0.45673
10	0.34155	0.65845	−0.4054	−0.4179	−0.8726	0.20824	0.49379
11	0.37676	0.62324	−0.3122	−0.4728	−0.7490	0.23827	0.52998
12	0.41197	0.58803	−0.2214	−0.5310	−0.6330	0.26911	0.56536
13	0.44718	0.55282	−0.1323	−0.5927	−0.5230	0.30072	0.59996
14	0.48239	0.51761	−0.0440	−0.6585	−0.4177	0.33309	0.63380
15	0.51761	0.48239	0.0440	−0.7290	−0.3161	0.36620	0.66691
16	0.55282	0.44718	0.1325	−0.8048	−0.2172	0.40004	0.69927
17	0.58803	0.41197	0.2221	−0.8868	−0.1201	0.43464	0.73089

Table T2 *Cont.*

i	F'(i)	1 − F'(i)		Xn(i)	Xe(i)	Xw(i)	p0.05	p0.95
18	0.62324	0.37676		0.3136	−0.9761	−0.0241	0.47002	0.46173
19	0.65845	0.34155		0.4078	−1.0743	0.0716	0.50621	0.79176
20	0.69366	0.30634		0.5058	−1.1831	0.1681	0.54327	0.82092
21	0.72887	0.27113		0.6091	−1.3052	0.2663	0.58127	0.84915
22	0.76408	0.23592		0.7192	−1.4443	0.3676	0.62033	0.87633
23	0.79930	0.20070		0.8389	−1.6059	0.4737	0.66060	0.90232
24	0.83451	0.16549		0.9721	−1.7988	0.5871	0.70231	0.92689
25	0.86972	0.13028		1.1251	−2.0381	0.7120	0.74583	0.94696
26	0.90493	0.09507		1.3104	−2.3531	0.8557	0.79179	0.97015
27	0.94014	0.05986		1.5563	−2.8158	1.0352	0.84149	0.98716
28	0.97535	0.02465		1.9665	−3.7031	1.3092	0.89853	0.99817
			S3:	0.3625	−27.0734	−15.4501		
			S4:	23.7000	48.2965	46.9491		

Total number of results: 29

i	F'(i)	1 − F'(i)		Xn(i)	Xe(i)	Xw(i)	p0.05	p0.95
1	0.02381	0.97619		−1.8144	−0.0241	−3.7256	0.00177	0.09814
2	0.05782	0.94218		−1.4964	−0.0596	−2.8207	0.01239	0.15339
3	0.09184	0.90816		−1.2842	−0.0963	−2.3400	0.02879	0.20156
4	0.12585	0.87415		−1.1170	−0.1345	−2.0062	0.04852	0.24614
5	0.15986	0.84014		−0.9752	−0.1742	−1.7476	0.07049	0.28837
6	0.19388	0.80612		−0.8500	−0.2155	−1.5347	0.09415	0.32887
7	0.22789	0.77211		−0.7362	−0.2586	−1.3524	0.11917	0.36800
8	0.26190	0.73810		−0.6307	−0.3037	−1.1918	0.14532	0.40597
9	0.29592	0.70408		−0.5314	−0.3509	−1.0474	0.17246	0.44294
10	0.32993	0.67007		−0.4369	−0.4004	−0.9154	0.20050	0.47901
11	0.36395	0.63605		−0.3458	−0.4525	−0.7930	0.22934	0.51427
12	0.39796	0.60204		−0.2573	−0.5074	−0.6784	0.25894	0.54877
13	0.43197	0.56803		−0.1706	−0.5656	−0.5699	0.28927	0.58254
14	0.46599	0.53401		−0.0851	−0.6273	−0.4663	0.32030	0.61561
15	0.50000	0.50000		0.0000	−0.6931	−0.3665	0.35200	0.64799
16	0.53401	0.46599		0.0852	−0.7636	−0.2697	0.38439	0.67970
17	0.56803	0.43197		0.1710	−0.8394	−0.1751	0.41746	0.71073
18	0.60204	0.39796		0.2582	−0.9214	−0.0819	0.45123	0.74106
19	0.63605	0.36395		0.3475	−1.0108	0.0107	0.48573	0.77066
20	0.67007	0.32993		0.4397	−1.1089	0.1033	0.52099	0.79950
21	0.70408	0.29592		0.5358	−1.2177	0.1969	0.55076	0.82753
22	0.73810	0.26190		0.6372	−1.3398	0.2925	0.59403	0.85468
23	0.77211	0.22789		0.7456	−1.4789	0.3913	0.63200	0.88083
24	0.80612	0.19388		0.8635	−1.6405	0.4950	0.67113	0.90584
25	0.84014	0.15986		0.9950	−1.8334	0.6062	0.71168	0.92951
26	0.87415	0.12585		1.1463	−2.0727	0.7288	0.75386	0.95148
27	0.90816	0.09184		1.3297	−2.3877	0.8703	0.79844	0.97120
28	0.94218	0.05782		1.5736	−2.8504	1.0474	0.84661	0.98761
29	0.97619	0.02381		1.9812	−3.7377	1.3185	0.90185	0.99823
			S3:	0.3783	−28.0665	−16.0214		
			S4:	24.6223	50.2044	48.8312		

Total number of results: 30

i	F'(i)	1 − F'(i)		Xn(i)	Xe(i)	Xw(i)	p0.05	p0.95
1	0.02303	0.97697		−1.8244	−0.0233	−3.7595	0.00171	0.09503
2	0.05592	0.94408		−1.5104	−0.0575	−2.8552	0.01189	0.14860

Table T2 *Cont.*

i	F'(i)	1 − F'(i)		Xn(i)	Xe(i)	Xw(i)	p0.05	p0.95
3	0.08882	0.91118		−1.3008	−0.0930	−2.3750	0.02781	0.19533
4	0.12171	0.87829		−1.1357	−0.1298	−2.0419	0.04685	0.23860
5	0.15461	0.84539		−0.9959	−0.1680	−1.7841	0.06806	0.27962
6	0.18750	0.81250		−0.8724	−0.2076	−1.5720	0.09087	0.31897
7	0.22039	0.77961		−0.7604	−0.2490	−1.3904	0.11499	0.35701
8	0.25329	0.74671		−0.6568	−0.2921	−1.2307	0.14018	0.39395
9	0.28618	0.71382		−0.5593	−0.3371	−1.0873	0.16633	0.42993
10	0.31908	0.68092		−0.4666	−0.3843	−0.9563	0.19331	0.46507
11	0.35197	0.64803		−0.3775	−0.4338	−0.8351	0.22106	0.49944
12	0.38487	0.61513		−0.2911	−0.4859	−0.7217	0.24953	0.53309
13	0.41776	0.58224		−0.2067	−0.5409	−0.6146	0.27867	0.56605
14	0.45066	0.54934		−0.1235	−0.5990	−0.5124	0.30846	0.59837
15	0.48355	0.51645		−0.0411	−0.6608	−0.4143	0.33889	0.63005
16	0.51645	0.48355		0.0411	−0.7266	−0.3194	0.36995	0.66111
17	0.54934	0.45066		0.1237	−0.7970	−0.2268	0.40163	0.69154
18	0.58224	0.41776		0.2072	−0.8728	−0.1360	0.43394	0.72133
19	0.61513	0.38487		0.2923	−0.9549	−0.0462	0.46691	0.75047
20	0.64803	0.35197		0.3796	−1.0442	0.0432	0.50056	0.77894
21	0.68092	0.31908		0.4699	−1.1423	0.1331	0.53493	0.80669
22	0.71382	0.28618		0.5642	−1.2511	0.2240	0.57007	0.83367
23	0.74671	0.25329		0.6639	−1.3732	0.3172	0.60605	0.85981
24	0.77961	0.22039		0.7706	−1.5123	0.4137	0.64299	0.88501
25	0.81250	0.18750		0.8870	−1.6740	0.5152	0.68103	0.90913
26	0.84539	0.15461		1.0169	−1.8669	0.6243	0.72038	0.93194
27	0.87829	0.12171		1.1666	−2.1061	0.7448	0.76140	0.95314
28	0.91118	0.08882		1.3483	−2.4212	0.8843	0.80467	0.97218
29	0.94408	0.05592		1.5903	−2.8838	1.0591	0.85140	0.98802
30	0.97697	0.02303		1.9953	−3.7711	1.3274	0.90497	0.99829
			S3:	0.3941	−29.0598	−16.5928		
			S4:	25.5451	52.1149	50.7161		

Total number of results: 31

i	F'(i)	1 − F'(i)	Xn(i)	Xe(i)	Xw(i)	p0.05	p0.95
1	0.02229	0.97771	−1.8340	−0.0225	−3.7922	0.00165	0.09211
2	0.05414	0.94586	−1.5237	−0.0557	−2.8885	0.01158	0.14409
3	0.08599	0.91401	−1.3167	−0.0899	−2.4089	0.02690	0.18946
4	0.11783	0.88217	−1.1536	−0.1254	−2.0764	0.04530	0.23150
5	0.14968	0.85032	−1.0156	−0.1621	−1.8193	0.06578	0.27137
6	0.18153	0.81847	−0.8939	−0.2003	−1.6079	0.08781	0.30964
7	0.21338	0.78662	−0.7836	−0.2400	−1.4271	0.11109	0.34665
8	0.24522	0.75478	−0.6815	−0.2813	−1.2682	0.13540	0.38261
9	0.27707	0.72293	−0.5858	−0.3244	−1.1256	0.16061	0.41766
10	0.30892	0.69108	−0.4948	−0.3695	−0.9956	0.18662	0.45190
11	0.34076	0.65924	−0.4075	−0.4167	−0.8755	0.21336	0.48542
12	0.37261	0.62739	−0.3230	−0.4662	−0.7632	0.24077	0.51825
13	0.40446	0.59554	−0.2406	−0.5183	−0.6572	0.26883	0.55044
14	0.43631	0.56369	−0.1597	−0.5732	−0.5564	0.29749	0.58203
15	0.46815	0.53185	−0.0796	−0.6314	−0.4598	0.32674	0.61302
16	0.50000	0.50000	0.0000	−0.6931	−0.3665	0.35657	0.64343
17	0.53185	0.46815	0.0797	−0.7590	−0.2758	0.38698	0.67326

Table T2 *Cont.*

i	F'(i)	1 − F'(i)	Xn(i)	Xe(i)	Xw(i)	p0.05	p0.95
18	0.56369	0.43631	0.1600	−0.8294	−0.1870	0.41797	0.70251
19	0.59554	0.40446	0.2414	−0.9052	−0.0996	0.44956	0.73117
20	0.62739	0.37261	0.3245	−0.9872	−0.0129	0.48175	0.75922
21	0.65924	0.34076	0.4099	−1.0766	0.0738	0.51458	0.78664
22	0.69108	0.30892	0.4985	−1.1747	0.1610	0.54810	0.81338
23	0.72293	0.27707	0.5912	−1.2835	0.2496	0.58234	0.83939
24	0.75478	0.24522	0.6893	−1.4056	0.3405	0.64739	0.86460
25	0.78662	0.21338	0.7946	−1.5447	0.4348	0.65336	0.88891
26	0.81847	0.18153	0.9094	−1.7063	0.5344	0.69036	0.91219
27	0.85032	0.14968	1.0378	−1.8992	0.6415	0.72563	0.93422
28	0.88217	0.11783	1.1860	−2.1385	0.7601	0.76650	0.95470
29	0.91401	0.08599	1.3661	−2.4536	0.8975	0.81054	0.97310
30	0.94586	0.05414	1.6063	−2.9162	1.0703	0.85591	0.98841
31	0.97771	0.02229	2.0090	−3.8035	1.3359	0.90789	0.99835
			S3: 0.4100	−30.0533	−17.1644		
			S4: 26.4685	54.0279	52.6035		

Total number of results: 32

i	F'(i)	1 − F'(i)	Xn(i)	Xe(i)	Xw(i)	p0.05	p0.95
1	0.02160	0.97840	−1.8432	−0.0218	−3.8239	0.00160	0.08937
2	0.05247	0.94753	−1.5365	−0.0539	−2.9207	0.01122	0.13985
3	0.08333	0.91667	−1.3319	−0.0870	−2.4417	0.02604	0.18394
4	0.11420	0.88580	−1.1708	−0.1213	−2.1098	0.04384	0.22482
5	0.14506	0.85494	−1.0345	−0.1567	−1.8533	0.06365	0.26360
6	0.17593	0.82407	−0.9144	−0.1935	−1.6425	0.08495	0.30084
7	0.20679	0.79321	−0.8056	−0.2317	−1.4625	0.10745	0.33687
8	0.23765	0.76235	−0.7051	−0.2714	−1.3043	0.13093	0.37190
9	0.26852	0.73148	−0.6110	−0.3127	−1.1626	0.15528	0.40606
10	0.29938	0.70062	−0.5216	−0.3558	−1.0334	0.18038	0.43945
11	0.33025	0.66975	−0.4360	−0.4008	−0.9142	0.20618	0.47214
12	0.36111	0.63889	−0.3533	−0.4480	−0.8029	0.23262	0.50419
13	0.39198	0.60802	−0.2727	−0.4975	−0.6981	0.25966	0.53564
14	0.42284	0.57716	−0.1938	−0.5496	−0.5985	0.28727	0.56651
15	0.45370	0.54630	−0.1159	−0.6046	−0.5032	0.31544	0.59683
16	0.48457	0.51543	−0.0386	−0.6627	−0.4114	0.34415	0.62661
17	0.51543	0.48457	0.0386	−0.7245	−0.3223	0.37339	0.65585
18	0.54630	0.45370	0.1160	−0.7903	−0.2353	0.40317	0.68456
19	0.57716	0.42284	0.1943	−0.8608	−0.1499	0.43349	0.71272
20	0.60802	0.39198	0.2737	−0.9366	−0.0655	0.46436	0.74034
21	0.63889	0.36111	0.3550	−1.0186	0.0184	0.49581	0.76738
22	0.66975	0.33025	0.4388	−1.1079	0.1025	0.52786	0.79382
23	0.70062	0.29938	0.5258	−1.2060	0.1873	0.56055	0.81961
24	0.73148	0.26852	0.6170	−1.3148	0.2737	0.59314	0.84472
25	0.76235	0.23765	0.7136	−1.4369	0.3625	0.62810	0.86907
26	0.79321	0.20679	0.8174	−1.5761	0.4549	0.66313	0.89255
27	0.82407	0.17593	0.9309	−1.7377	0.5526	0.69916	0.91505
28	0.85494	0.14506	1.0579	−1.9306	0.6578	0.73640	0.93635
29	0.88580	0.11420	1.2046	−2.1698	0.7746	0.77518	0.95615
30	0.91667	0.08333	1.3832	−2.4849	0.9102	0.81606	0.97396
31	0.94753	0.05247	1.6217	−2.9475	1.0810	0.86015	0.98878

Table T2 *Cont.*

i	$F'(i)$	$1 - F'(i)$		Xn(i)	Xe(i)	Xw(i)	p0.05	p0.95
32	0.97840	0.02160		2.0221	3.8348	1.3441	0.91063	0.99840
			S3:	0.4259	−31.0470	−17.7363		
			S4:	27.3923	55.9432	54.4935		

Total number of results: 33

i	$F'(i)$	$1 - F'(i)$		Xn(i)	Xe(i)	Xw(i)	p0.05	p0.95
1	0.02096	0.97904		−1.8520	−0.0212	−3.8547	0.00155	0.08678
2	0.05090	0.94910		−1.5488	−0.0522	−2.9519	0.01086	0.13585
3	0.08084	0.91916		−1.3464	−0.0843	−2.4735	0.02524	0.17873
4	0.11078	0.88922		−1.1872	−0.1174	−2.1421	0.04246	0.21850
5	0.14072	0.85928		−1.0525	−0.1517	−1.8861	0.06166	0.25625
6	0.17066	0.82934		−0.9340	−0.1871	−1.6760	0.08227	0.29252
7	0.20060	0.79940		−0.8267	−0.2239	−1.4966	0.10404	0.32763
8	0.23054	0.76946		−0.7277	−0.2621	−1.3392	0.12675	0.36176
9	0.26048	0.73952		−0.6350	−0.3018	−1.1981	0.15029	0.39507
10	0.29042	0.70958		−0.5471	−0.3431	−1.0698	0.17455	0.42765
11	0.32036	0.67964		−0.4631	−0.3862	−0.9514	0.19948	0.45956
12	0.35030	0.64970		−0.3820	−0.4312	−0.8411	0.22501	0.49086
13	0.38024	0.61976		−0.3031	−0.4784	−0.7373	0.25111	0.52159
14	0.41018	0.58982		−0.2260	−0.5279	−0.6388	0.27775	0.55177
15	0.44012	0.55988		−0.1500	−0.5800	−0.5447	0.30491	0.58144
16	0.47006	0.52994		−0.0749	−0.6350	−0.4541	0.33258	0.61060
17	0.50000	0.50000		0.0000	−0.6931	−0.3665	0.36074	0.63926
18	0.52994	0.47006		0.0749	−0.7549	−0.2812	0.38940	0.66742
19	0.55988	0.44012		0.1503	−0.8207	−0.1976	0.41656	0.69509
20	0.58982	0.41018		0.2267	−0.8912	−0.1152	0.44823	0.72225
21	0.61976	0.38024		0.3044	−0.9670	−0.0336	0.47841	0.74889
22	0.64970	0.35030		0.3841	−1.0490	0.0478	0.50914	0.77499
23	0.67964	0.32036		0.4663	−1.1383	0.1295	0.54344	0.80052
24	0.70958	0.29042		0.5518	−1.2364	0.2122	0.57235	0.82545
25	0.73952	0.26048		0.6415	−1.3452	0.2966	0.60493	0.84971
26	0.76946	0.23054		0.7368	−1.4673	0.3834	0.63824	0.87325
27	0.79940	0.20060		0.8393	−1.6064	0.4740	0.67237	0.89596
28	0.82934	0.17066		0.9515	−1.7681	0.5699	0.70748	0.91772
29	0.85928	0.14072		1.0771	−1.9610	0.6735	0.74375	0.93834
30	0.88922	0.11078		1.2225	−2.2002	0.7886	0.76150	0.95752
31	0.91916	0.08084		1.3997	−2.5153	0.9224	0.82127	0.97476
32	0.94910	0.05090		1.6366	−2.9779	1.0912	0.86415	0.98912
33	0.97904	0.02096		2.0348	−3.8652	1.3520	0.91322	0.99845
			S3:	0.4441	−32.0409	−18.3082		
			S4:	28.3166	57.8607	56.3858		

Total number of results: 34

i	$F'(i)$	$1 - F'(i)$		Xn(i)	Xe(i)	Xw(i)	p0.05	p0.95
1	0.02035	0.97965		−1.8605	−0.0206	−3.8845	0.00151	0.08434
2	0.04942	0.95058		−1.5606	−0.0507	−2.9822	0.01055	0.13207
3	0.07849	0.92151		−1.3604	−0.0817	−2.5042	0.02448	0.17381
4	0.10756	0.89244		−1.2029	−0.1138	−2.1734	0.04120	0.21253
5	0.13663	0.86337		−1.0698	−0.1469	−1.9179	0.05978	0.24931
6	0.16570	0.83430		−0.9527	−0.1812	−1.7084	0.07976	0.28465
7	0.19477	0.80523		−0.8468	−0.2166	−1.5296	0.10084	0.31887
8	0.22384	0.77616		−0.7492	−0.2534	−1.3728	0.12283	0.35216

Table T2 *Cont.*

i	F'(i)	1 − F'(i)	Xn(i)	Xe(i)	Xw(i)	p0.05	p0.95
9	0.25291	0.74709	−0.6579	−0.2916	−1.2325	0.14561	0.38466
10	0.28198	0.71802	−0.5715	−0.3313	−1.1049	0.16909	0.41645
11	0.31105	0.68895	−0.4889	−0.3726	−0.9873	0.19319	0.44761
12	0.34012	0.65988	−0.4093	−0.4157	−0.8778	0.21788	0.47819
13	0.36919	0.63081	−0.3320	−0.4607	−0.7749	0.24310	0.50823
14	0.39826	0.60174	−0.2565	−0.5079	−0.6774	0.26884	0.53775
15	0.42733	0.57267	−0.1824	−0.5574	−0.5844	0.29507	0.56678
16	0.45640	0.54360	−0.1091	−0.6095	−0.4951	0.32177	0.59534
17	0.48547	0.51453	−0.0363	−0.6645	−0.4087	0.34894	0.62343
18	0.51453	0.48547	0.0363	−0.7226	−0.3248	0.37657	0.65106
19	0.54360	0.45640	0.1093	−0.7844	−0.2428	0.40466	0.67823
20	0.57267	0.42733	0.1828	−0.8502	−0.1623	0.43321	0.70493
21	0.60174	0.39826	0.2574	−0.9207	−0.0827	0.46225	0.73116
22	0.63081	0.36919	0.3336	−0.9965	−0.0036	0.49177	0.75689
23	0.65988	0.34012	0.4117	−1.0785	0.0755	0.52181	0.78212
24	0.68895	0.31105	0.4925	−1.1678	0.1551	0.55239	0.80680
25	0.71802	0.28198	0.5766	−1.2659	0.2358	0.58355	0.83091
26	0.74709	0.25291	0.6651	−1.3747	0.3183	0.61534	0.85439
27	0.77616	0.22384	0.7591	−1.4968	0.4034	0.64754	0.87717
28	0.80523	0.19477	0.8603	−1.6359	0.4922	0.68113	0.89916
29	0.83430	0.16570	0.9713	−1.7976	0.5864	0.71535	0.92024
30	0.86337	0.13663	1.0956	−1.9905	0.6884	0.75069	0.94021
31	0.89244	0.10756	1.2398	−2.2297	0.8019	0.78747	0.95880
32	0.92151	0.07849	1.4156	−2.5448	0.9341	0.82619	0.97552
33	0.95058	0.04942	1.6509	−3.0074	1.1011	0.86793	0.98945
34	0.97965	0.02035	2.0470	−3.8947	1.3596	0.91566	0.99849
			S3: 0.4578	−33.0349	−18.8804		
			S4: 29.2413	59.7803	58.2802		

Total number of results: 35

i	F'(i)	1 − F'(i)	Xn(i)	Xe(i)	Xw(i)	p0.05	p0.95
1	0.01977	0.98023	−1.8686	−0.0200	−3.9134	0.00146	0.08203
2	0.04802	0.95198	−1.5719	−0.0492	−3.0116	0.01025	0.12850
3	0.07627	0.92373	−1.3738	−0.0793	−2.5341	0.02377	0.16915
4	0.10452	0.89548	−1.2180	−0.1104	−2.2037	0.03999	0.20688
5	0.13277	0.86723	−1.0865	−0.1424	−1.9488	0.05802	0.24272
6	0.16102	0.83898	−0.9707	−0.1756	−1.7397	0.07739	0.27718
7	0.18927	0.81073	−0.8662	−0.2098	−1.5615	0.09783	0.31056
8	0.21751	0.78249	−0.7699	−0.2453	−1.4054	0.11914	0.34305
9	0.24576	0.75424	−0.6799	−0.2820	−1.2657	0.14122	0.37477
10	0.27401	0.72599	−0.5948	−0.3202	−1.1387	0.16396	0.40582
11	0.30226	0.69774	−0.5135	−0.3599	−1.0219	0.18730	0.43626
12	0.33051	0.66949	−0.4353	−0.4012	−0.9132	0.21119	0.46615
13	0.35876	0.64124	−0.3595	−0.4443	−0.8112	0.23560	0.49552
14	0.38701	0.61299	−0.2856	−0.4894	−0.7146	0.26049	0.52440
15	0.41525	0.58475	−0.2130	−0.5366	−0.6225	0.28585	0.55282
16	0.44350	0.55650	−0.1415	−0.5861	−0.5343	0.31165	0.58080
17	0.47175	0.52825	−0.0706	−0.6382	−0.4491	0.33789	0.60833
18	0.50000	0.50000	0.0000	−0.6931	−0.3665	0.36457	0.63543
19	0.52825	0.47175	0.0707	−0.7513	−0.2859	0.39167	0.66210

Table T2 *Cont.*

i	F'(i)	1 − F'(i)		Xn(i)	Xe(i)	Xw(i)	p0.05	p0.95
20	0.55650	0.44350		0.1418	−0.8131	−0.2070	0.41920	0.68835
21	0.58475	0.41525		0.2137	−0.8789	−0.1291	0.44717	0.71415
22	0.61299	0.38701		0.2867	−0.9493	−0.0520	0.47560	0.73951
23	0.64124	0.35876		0.3613	−1.0251	0.0248	0.50448	0.76440
24	0.66949	0.33051		0.4381	−1.1071	0.1018	0.53385	0.78881
25	0.69774	0.30226		0.5175	−1.1965	0.1794	0.56374	0.81270
26	0.72599	0.27401		0.6004	−1.2946	0.2582	0.59416	0.83604
27	0.75424	0.24576		0.6876	−1.4034	0.3389	0.62523	0.85878
28	0.78249	0.21751		0.7804	−1.5255	0.4223	0.65695	0.88086
29	0.81073	0.18927		0.8805	−1.6646	0.5096	0.68944	0.90217
30	0.83898	0.16102		0.9902	−1.8262	0.6023	0.72282	0.92261
31	0.86723	0.13277		1.1135	−2.0191	0.7027	0.75728	0.94198
32	0.89548	0.10452		1.2564	−2.2584	0.8146	0.79312	0.96001
33	0.92373	0.07627		1.4309	−2.5735	0.9453	0.83085	0.97623
34	0.95198	0.04802		1.6647	−3.0361	1.1106	0.87150	0.98975
35	0.98023	0.01977		2.0589	−3.9234	1.3670	0.91797	0.99854
			S3:	0.4739	−34.0292	−19.4526		
			S4:	30.1665	61.7019	60.1767		

Total number of results: 36

i	F'(i)	1 − F'(i)	Xn(i)	Xe(i)	Xw(i)	p0.05	p0.95
1	0.01923	0.98077	−1.8764	−0.0194	−3.9416	0.00142	0.07985
2	0.04670	0.95330	−1.5828	−0.0478	−3.0401	0.00996	0.12512
3	0.07418	0.92582	−1.3867	−0.0771	−2.5630	0.02310	0.16474
4	0.10165	0.89835	−1.2326	−0.1072	−2.2331	0.03885	0.20152
5	0.12912	0.87088	−1.1024	−0.1383	−1.9787	0.05636	0.23648
6	0.15659	0.84341	−0.9880	−0.1703	−1.7702	0.07516	0.27010
7	0.18407	0.81593	−0.8847	−0.2034	−1.5925	0.09499	0.30268
8	0.21154	0.78846	−0.7897	−0.2377	−1.4369	0.11567	0.33439
9	0.23901	0.76099	−0.7009	−0.2731	−1.2978	0.13708	0.36537
10	0.26648	0.73352	−0.6170	−0.3099	−1.1715	0.15913	0.39571
11	0.29396	0.70604	−0.5370	−0.3481	−1.0553	0.18175	0.42546
12	0.32143	0.67857	−0.4601	−0.3878	−0.9474	0.20491	0.45468
13	0.34890	0.65110	−0.3857	−0.4291	−0.8461	0.22855	0.48341
14	0.37637	0.62363	−0.3132	−0.4722	−0.7503	0.25265	0.51168
15	0.40385	0.59615	−0.2422	−0.5173	−0.6592	0.27719	0.53951
16	0.43132	0.56868	−0.1723	−0.5644	−0.5719	0.30216	0.56691
17	0.45879	0.54121	−0.1031	−0.6139	−0.4878	0.32754	0.59391
18	0.48626	0.51374	−0.0343	−0.6660	−0.4064	0.35332	0.62049
19	0.51374	0.48626	0.0343	−0.7210	−0.3271	0.37951	0.64668
20	0.54121	0.45879	0.1032	−0.7792	−0.2495	0.40609	0.67246
21	0.56868	0.43132	0.1727	−0.8409	−0.1733	0.43309	0.69784
22	0.59615	0.40385	0.2430	−0.9067	−0.0979	0.46049	0.72280
23	0.62363	0.37637	0.3146	−0.9772	−0.0231	0.48832	0.74735
24	0.65110	0.34890	0.3878	−1.0530	0.0516	0.51658	0.77145
25	0.67857	0.32143	0.4633	−1.1350	0.1266	0.54532	0.79509
26	0.70604	0.29396	0.5415	−1.2243	0.2024	0.57454	0.81825
27	0.73352	0.26648	0.6231	−1.3224	0.2795	0.60429	0.84087
28	0.76099	0.23901	0.7092	−1.4312	0.3585	0.63483	0.86292
29	0.78846	0.21154	0.8009	−1.5533	0.4404	0.66561	0.88433

Table T2 *Cont.*

i	F'(i)	1 − F'(i)	Xn(i)	Xe(i)	Xw(i)	p0.05	p0.95
30	0.81593	0.18407	0.8999	−1.6925	0.5262	0.69732	0.90501
31	0.84341	0.15659	1.0085	−1.8541	0.6174	0.72990	0.92483
32	0.87088	0.12912	1.1306	−2.0470	0.7164	0.76352	0.94364
33	0.89835	0.10165	1.2724	−2.2862	0.8269	0.79848	0.96114
34	0.92582	0.07418	1.4457	−2.6013	0.9560	0.83526	0.97690
35	0.95330	0.04670	1.6781	−3.0639	1.1197	0.87488	0.99004
36	0.98077	0.01923	2.0703	−3.9512	1.3740	0.92015	0.99858
			S3: 0.4899	−35.0236	−20.0250		
			S4: 31.0920	63.6253	62.0751		

Total number of results: 37

i	F'(i)	1 − F'(i)	Xn(i)	Xe(i)	Xw(i)	p0.05	p0.95
1	0.01872	0.98128	−1.8839	−0.0189	−3.9689	0.00138	0.07778
2	0.04545	0.95455	−1.5932	−0.0465	−3.0679	0.00969	0.12191
3	0.07219	0.92781	−1.3992	−0.0749	−2.5912	0.02246	0.16054
4	0.09893	0.90107	−1.2466	−0.1042	−2.2617	0.03778	0.19643
5	0.12567	0.87433	−1.1178	−0.1343	−2.0077	0.05479	0.23054
6	0.15241	0.84759	−1.0047	−0.1654	−1.7997	0.07306	0.26337
7	0.17914	0.82086	−0.9026	−0.1974	−1.6225	0.09232	0.29518
8	0.20588	0.79412	−0.8087	−0.2305	−1.4674	0.11240	0.32616
9	0.23262	0.76738	−0.7211	−0.2648	−1.3289	0.13318	0.35641
10	0.25936	0.74064	−0.6384	−0.3002	−1.2032	0.15458	0.38608
11	0.28610	0.71390	−0.5596	−0.3370	−1.0877	0.17653	0.41517
12	0.31283	0.68717	−0.4839	−0.3752	−0.9804	0.19898	0.44376
13	0.33957	0.66043	−0.4107	−0.4149	−0.8798	0.22191	0.47187
14	0.36631	0.63369	−0.3396	−0.4562	−0.7848	0.24527	0.49955
15	0.39305	0.60695	−0.2700	−0.4993	−0.6945	0.26905	0.52680
16	0.41979	0.58021	−0.2015	−0.5444	−0.6081	0.29324	0.55366
17	0.44652	0.55348	−0.1339	−0.5915	−0.5250	0.31781	0.58012
18	0.47326	0.52674	−0.0668	−0.6411	−0.4446	0.34276	0.60620
19	0.50000	0.50000	0.0000	−0.6931	−0.3665	0.36809	0.63190
20	0.52674	0.47326	0.0669	−0.7481	−0.2902	0.39380	0.65723
21	0.55348	0.44652	0.1342	−0.8063	−0.2153	0.41988	0.68219
22	0.58021	0.41979	0.2021	−0.8680	−0.1416	0.44634	0.70676
23	0.60695	0.39305	0.2710	−0.9338	−0.0685	0.47320	0.73094
24	0.63369	0.36631	0.3412	−1.0043	0.0043	0.50045	0.75473
25	0.66043	0.33957	0.4132	−1.0801	0.0770	0.52812	0.77809
26	0.68717	0.31283	0.4874	−1.1621	0.1502	0.55624	0.80101
27	0.71390	0.28610	0.5644	−1.2514	0.2243	0.58483	0.82347
28	0.74064	0.25936	0.6450	−1.3495	0.2998	0.61392	0.84542
29	0.76738	0.23262	0.7300	−1.4583	0.3773	0.64357	0.86682
30	0.79412	0.20588	0.8206	−1.5805	0.4577	0.67384	0.88760
31	0.82086	0.17914	0.9185	−1.7196	0.5421	0.70482	0.90768
32	0.84759	0.15241	1.0262	−1.8812	0.6319	0.73663	0.92694
33	0.87433	0.12567	1.1472	−2.0741	0.7295	0.76946	0.94521
34	0.90107	0.09893	1.2879	−2.3133	0.8387	0.80357	0.96222
35	0.92781	0.07219	1.4599	−2.6284	0.9664	0.83946	0.97754
36	0.95455	0.04545	1.6910	−3.0910	1.1285	0.87809	0.99031
37	0.98128	0.01872	2.0814	−3.9783	1.3809	0.92222	0.99861
			S3: 0.5061	−36.0182	−20.5976		
			S4: 32.0178	65.5505	63.9755		

Table T2 *Cont.*

i	F'(i)	1 − F'(i)	Xn(i)	Xe(i)	Xw(i)	p0.05	p0.95
	Total number of results: 38						
1	0.01823	0.98177	−1.8911	−0.0184	−3.9956	0.00135	0.07581
2	0.04427	0.95573	−1.6033	−0.0453	−3.0949	0.00943	0.11885
3	0.07031	0.92969	−1.4111	−0.0729	−2.6186	0.02186	0.15656
4	0.09635	0.90365	−1.2600	−0.1013	−2.2895	0.03676	0.19159
5	0.12240	0.87760	−1.1326	−0.1306	−2.0359	0.05331	0.22490
6	0.14844	0.85156	−1.0207	−0.1607	−1.8283	0.07107	0.25696
7	0.17448	0.82552	−0.9197	−0.1917	−1.6516	0.08979	0.28804
8	0.20052	0.79948	−0.8269	−0.2238	−1.4970	0.10931	0.31832
9	0.22656	0.77344	−0.7404	−0.2569	−1.3590	0.12950	0.34791
10	0.25260	0.74740	−0.6588	−0.2912	−1.2339	0.15028	0.37691
11	0.27865	0.72135	−0.5812	−0.3266	−1.1189	0.17160	0.40537
12	0.30469	0.69531	−0.5067	−0.3634	−1.0123	0.19340	0.43334
13	0.33073	0.66927	−0.4347	−0.4016	−0.9124	0.21565	0.46086
14	0.35677	0.64323	−0.3648	−0.4413	−0.8181	0.23832	0.48796
15	0.38281	0.61719	−0.2964	−0.4826	−0.7286	0.26178	0.51466
16	0.40885	0.59115	−0.2294	−0.5257	−0.6430	0.28483	0.54098
17	0.43490	0.56510	−0.1632	−0.5707	−0.5608	0.30865	0.56693
18	0.46094	0.53906	−0.0977	−0.6179	−0.4814	0.33283	0.59252
19	0.48698	0.51302	−0.0325	−0.6674	−0.4043	0.35736	0.61776
20	0.51302	0.48698	0.0326	−0.7195	−0.3292	0.38224	0.64264
21	0.53906	0.46094	0.0978	−0.7745	−0.2555	0.40748	0.66717
22	0.56510	0.43490	0.1636	−0.8326	−0.1831	0.43307	0.69135
23	0.59115	0.40885	0.2301	−0.8944	−0.1116	0.45902	0.71517
24	0.61719	0.38281	0.2977	−0.9602	−0.0406	0.48534	0.73862
25	0.64323	0.35677	0.3667	−1.0307	0.0302	0.51204	0.76168
26	0.66927	0.33073	0.4375	−1.1065	0.1012	0.53914	0.78435
27	0.69531	0.30469	0.5106	−1.1885	0.1727	0.56666	0.80660
28	0.72135	0.27865	0.5865	−1.2778	0.2452	0.59463	0.82840
29	0.74740	0.25260	0.6660	−1.3759	0.3191	0.62309	0.84972
30	0.77344	0.22656	0.7500	−1.4847	0.3952	0.65209	0.87050
31	0.79948	0.20052	0.8396	−1.6068	0.4743	0.68168	0.89069
32	0.82552	0.17448	0.9365	−1.7459	0.5573	0.71196	0.91021
33	0.85156	0.14844	1.0432	−1.9076	0.6458	0.74304	0.92893
34	0.87760	0.12240	1.1632	−2.1005	0.7422	0.77510	0.94669
35	0.90365	0.09635	1.3028	−2.3397	0.8500	0.80841	0.96324
36	0.92969	0.07031	1.4738	−2.6548	0.9764	0.84344	0.97814
37	0.95573	0.04427	1.7035	−3.1174	1.1370	0.88115	0.99057
38	0.98177	0.01823	2.0922	−4.0047	1.3875	0.92419	0.99865
		S3:	0.5222	−37.0129	−21.1702		
		S4:	32.9440	67.4774	65.8776		
	Total number of results: 39						
1	0.01777	0.98223	−1.8981	−0.0179	−4.0215	0.00131	0.07394
2	0.04315	0.95685	−1.6131	−0.0441	−3.1212	0.00919	0.11595
3	0.06853	0.93147	−1.4227	−0.0710	−2.6452	0.02129	0.15277
4	0.09391	0.90609	−1.2731	−0.0986	−2.3165	0.03580	0.18698
5	0.11929	0.88071	−1.1469	−0.1270	−2.0634	0.05190	0.21952
6	0.14467	0.85533	−1.0361	−0.1563	−1.8562	0.06919	0.25085

Table T2 *Cont.*

i	$F'(i)$	$1 - F'(i)$	Xn(i)	Xe(i)	Xw(i)	p0.05	p0.95
7	0.17005	0.82995	−0.9363	−0.1864	−1.6799	0.08740	0.28124
8	0.19543	0.80457	−0.8445	−0.2174	−1.5258	0.10638	0.31084
9	0.22081	0.77919	−0.7591	−0.2495	−1.3883	0.12601	0.33979
10	0.24619	0.75381	−0.6785	−0.2826	−1.2637	0.14622	0.36815
11	0.27157	0.72843	−0.6019	−0.3169	−1.1493	0.16694	0.39601
12	0.29695	0.70305	−0.5285	−0.3523	−1.0432	0.18812	0.42339
13	0.32234	0.67767	−0.4576	−0.3891	−0.9439	0.20973	0.45034
14	0.34772	0.65228	−0.3889	−0.4273	−0.8503	0.23175	0.47689
15	0.37310	0.62690	−0.3218	−0.4670	−0.7615	0.25414	0.50305
16	0.39848	0.60152	−0.2560	−0.5083	−0.6767	0.27690	0.52886
17	0.42386	0.57614	−0.1912	−0.5514	−0.5953	0.30001	0.55431
18	0.44924	0.55076	−0.1271	−0.5965	−0.5168	0.32346	0.57942
19	0.47462	0.52538	−0.0634	−0.6436	−0.4406	0.34725	0.60419
20	0.50000	0.50000	0.0000	−0.6931	−0.3665	0.37136	0.62864
21	0.52538	0.47462	0.0635	−0.7452	−0.2940	0.39581	0.65275
22	0.55076	0.44924	0.1273	−0.8002	−0.2229	0.42058	0.67654
23	0.57614	0.42386	0.1917	−0.8584	−0.1527	0.44569	0.69999
24	0.60152	0.39848	0.2569	−0.9201	−0.0833	0.47114	0.72310
25	0.62690	0.37310	0.3232	−0.9859	−0.0142	0.49694	0.74586
26	0.65228	0.34772	0.3911	−1.0564	0.0548	0.52311	0.76825
27	0.67767	0.32234	0.4608	−1.1322	0.1241	0.54966	0.79027
28	0.70305	0.29695	0.5328	−1.2142	0.1941	0.57661	0.81188
29	0.72843	0.27157	0.6077	−1.3035	0.2651	0.60399	0.83306
30	0.75381	0.24619	0.6862	−1.4016	0.3376	0.63185	0.85378
31	0.77919	0.22081	0.7692	−1.5104	0.4124	0.66021	0.87399
32	0.80457	0.19543	0.8579	−1.6325	0.4901	0.68916	0.89362
33	0.82995	0.17005	0.9539	−1.7717	0.5719	0.71876	0.91260
34	0.85533	0.14467	1.0596	−1.9333	0.6592	0.74915	0.93081
35	0.88071	0.11929	1.1787	−2.1262	0.7543	0.78048	0.94810
36	0.90609	0.09391	1.3173	−2.3654	0.8610	0.81302	0.96420
37	0.93147	0.06853	1.4871	−2.6805	0.9860	0.84723	0.97871
38	0.95685	0.04315	1.7157	−3.1431	1.1452	0.88405	0.99081
39	0.98223	0.01777	2.1027	−4.0304	1.3939	0.92606	0.99869
			S3: 0.5384	−38.0077	−21.7430		
			S4: 33.8705	69.4059	67.7814		

Total number of results: 40

i	$F'(i)$	$1 - F'(i)$	Xn(i)	Xe(i)	Xw(i)	p0.05	p0.95
1	0.01733	0.98267	−1.9048	−0.0175	−4.0468	0.00128	0.07216
2	0.04208	0.95792	−1.6225	−0.0430	−3.1468	0.00896	0.11319
3	0.06683	0.93317	−1.4339	−0.0692	−2.6712	0.02079	0.14915
4	0.09158	0.90842	−1.2856	−0.0961	−2.3429	0.03488	0.18259
5	0.11634	0.88366	−1.1606	−0.1237	−2.0901	0.05057	0.21440
6	0.14109	0.85891	−1.0510	−0.1521	−1.8833	0.06740	0.24503
7	0.16584	0.83416	−0.9522	−0.1813	−1.7074	0.08513	0.27475
8	0.19059	0.80941	−0.8615	−0.2115	−1.5537	0.10361	0.30371
9	0.21535	0.78465	−0.7770	−0.2425	−1.4167	0.12271	0.33203
10	0.24010	0.75990	−0.6975	−0.2746	−1.2926	0.14237	0.35979
11	0.26485	0.73515	−0.6219	−0.3077	−1.1787	0.16252	0.38706
12	0.28960	0.71040	−0.5495	−0.3419	−1.0731	0.18312	0.41388

Table T2 *Cont.*

i	F'(i)	1 − F'(i)	Xn(i)	Xe(i)	Xw(i)	p0.05	p0.95
13	0.31436	0.68564	−0.4797	−0.3774	−0.9745	0.20413	0.44028
14	0.33911	0.66089	−0.4120	−0.4142	−0.8815	0.22553	0.46630
15	0.36386	0.63614	−0.3460	−0.4523	−0.7933	0.24729	0.49195
16	0.38861	0.61139	−0.2814	−0.4920	−0.7092	0.26941	0.51725
17	0.41337	0.58663	−0.2179	−0.5334	−0.6286	0.29189	0.54222
18	0.43812	0.56188	−0.1551	−0.5765	−0.5508	0.31461	0.56686
19	0.46287	0.53713	−0.0929	−0.6215	−0.4756	0.33770	0.59119
20	0.48762	0.51238	−0.0309	−0.6687	−0.4024	0.36105	0.61520
21	0.51238	0.48762	0.0309	−0.7182	−0.3310	0.38480	0.63891
22	0.53713	0.46287	0.0930	−0.7703	−0.2610	0.40881	0.66230
23	0.56188	0.43812	0.1554	−0.8253	−0.1921	0.43314	0.68539
24	0.58663	0.41337	0.2185	−0.8834	−0.1240	0.45778	0.70815
25	0.61139	0.38861	0.2825	−0.9452	−0.0564	0.48275	0.73060
26	0.63614	0.36386	0.3477	−1.0110	0.0109	0.50805	0.75270
27	0.66089	0.33911	0.4145	−1.0814	0.0783	0.53370	0.77447
28	0.68564	0.31436	0.4831	−1.1572	0.1460	0.55972	0.79587
29	0.71040	0.28960	0.5542	−1.2392	0.2145	0.58612	0.81688
30	0.73515	0.26485	0.6281	−1.3286	0.2841	0.61294	0.83746
31	0.75990	0.24010	0.7057	−1.4267	0.3554	0.64021	0.85763
32	0.78465	0.21535	0.7878	−1.5355	0.4289	0.66797	0.87729
33	0.80941	0.19059	0.8756	−1.6576	0.5054	0.69629	0.89639
34	0.83416	0.16584	0.9707	−1.7967	0.5860	0.72525	0.91487
35	0.85891	0.14109	1.0755	−1.9584	0.6721	0.75497	0.93260
36	0.88366	0.11634	1.1936	−2.1513	0.7661	0.78560	0.94943
37	0.90842	0.09158	1.3313	−2.3905	0.8715	0.81741	0.96511
38	0.93317	0.06683	1.5001	−2.7056	0.9953	0.85085	0.97925
39	0.95792	0.04208	1.7274	−3.1682	1.1532	0.88681	0.99104
40	0.98267	0.01733	2.1128	−4.0555	1.4001	0.92784	0.99872
			S3: 0.5545	−39.0027	−22.3159		
			S4: 34.7973	71.3360	69.6868		

Total number of results: 41

i	F'(i)	1 − F'(i)	Xn(i)	Xe(i)	Xw(i)	p0.05	p0.95
1	0.01691	0.98309	−1.9113	−0.0171	−4.0715	0.00125	0.07046
2	0.04106	0.95894	−1.6316	−0.0419	−3.1718	0.00874	0.11055
3	0.06522	0.93478	−1.4447	−0.0674	−2.6965	0.02024	0.14571
4	0.08937	0.91063	−1.2978	−0.0936	−2.3685	0.03402	0.17840
5	0.11353	0.88647	−1.1740	−0.1205	−2.1161	0.04930	0.20951
6	0.13768	0.86232	−1.0654	−0.1481	−1.9097	0.06570	0.23947
7	0.16184	0.83816	−0.9676	−0.1765	−1.7342	0.08298	0.26854
8	0.18599	0.81401	−0.8778	−0.2058	−1.5809	0.10097	0.29689
9	0.21014	0.78986	−0.7943	−0.2359	−1.4443	0.11958	0.32461
10	0.23430	0.76570	−0.7157	−0.2670	−1.3206	0.13872	0.35180
11	0.25845	0.74155	−0.6411	−0.2990	−1.2073	0.15833	0.37851
12	0.28261	0.71739	−0.5697	−0.3321	−1.1022	0.17838	0.40478
13	0.30676	0.69324	−0.5008	−0.3664	−1.0041	0.19883	0.43065
14	0.33092	0.66908	−0.4342	−0.4018	−0.9117	0.21964	0.45615
15	0.35507	0.64493	−0.3693	−0.4386	−0.8241	0.24081	0.48131
16	0.37923	0.62077	−0.3058	−0.4768	−0.7407	0.26230	0.50612
17	0.40338	0.59662	−0.2434	−0.5165	−0.6607	0.28412	0.53062

Table T2 *Cont.*

i	F'(i)	1 − F'(i)	Xn(i)	Xe(i)	Xw(i)	p0.05	p0.95
18	0.42754	0.57246	−0.1818	−0.5578	−0.5837	0.30624	0.55482
19	0.45169	0.54831	−0.1209	−0.6009	−0.5093	0.32867	0.57871
20	0.47585	0.52415	−0.0604	−0.6460	−0.4370	0.35138	0.60230
21	0.50000	0.50000	0.0000	−0.6931	−0.3665	0.37440	0.62560
22	0.52415	0.47585	0.0604	−0.7427	−0.2975	0.39770	0.64861
23	0.54831	0.45169	0.1211	−0.7948	−0.2297	0.42129	0.67133
24	0.57246	0.42754	0.1823	−0.8497	−0.1629	0.44518	0.69376
25	0.59662	0.40338	0.2442	−0.9079	−0.0967	0.46937	0.71588
26	0.62077	0.37923	0.3071	−0.9696	−0.0309	0.49388	0.73769
27	0.64493	0.35507	0.3712	−1.0354	0.0348	0.51869	0.75919
28	0.66908	0.33092	0.4369	−1.1059	0.1006	0.54385	0.78035
29	0.69324	0.30676	0.5046	−1.1817	0.1669	0.56935	0.80117
30	0.71739	0.28261	0.5747	−1.2637	0.2340	0.59522	0.82162
31	0.74155	0.25845	0.6478	−1.3530	0.3024	0.62149	0.84166
32	0.76570	0.23430	0.7245	−1.4512	0.3724	0.64820	0.86128
33	0.78986	0.21014	0.8057	−1.5600	0.4447	0.67539	0.88042
34	0.81401	0.18599	0.8926	−1.6821	0.5200	0.70311	0.89903
35	0.83816	0.16184	0.9869	−1.8212	0.5995	0.73146	0.91702
36	0.86232	0.13768	1.0908	−1.9828	0.6845	0.76053	0.93430
37	0.88647	0.11353	1.2081	−2.1757	0.7774	0.79049	0.95070
38	0.91063	0.08937	1.3448	−2.4149	0.8817	0.82160	0.96598
39	0.93478	0.06522	1.5127	−2.7300	1.0043	0.85429	0.97976
40	0.95894	0.04106	1.7389	−3.1927	1.1609	0.88945	0.99126
41	0.98309	0.01691	2.1227	−4.0800	1.4061	0.92954	0.99875
S3:			0.5708	−39.9978	−22.8888		
S4:			35.7245	73.2675	71.5937		

Total number of results: 42

i	F'(i)	1 − F'(i)	Xn(i)	Xe(i)	Xw(i)	p0.05	p0.95
1	0.01651	0.98349	−1.9176	−0.0166	−4.0955	0.00122	0.06884
2	0.04009	0.95991	−1.6404	−0.0409	−3.1961	0.00853	0.10804
3	0.06368	0.93632	−1.4551	−0.0658	−2.7212	0.01975	0.14241
4	0.08726	0.91274	−1.3095	−0.0913	−2.3935	0.03319	0.17439
5	0.11085	0.88915	−1.1868	−0.1175	−2.1414	0.04810	0.20483
6	0.13443	0.86557	−1.0793	−0.1444	−1.9354	0.06409	0.23416
7	0.15802	0.84198	−0.9824	−0.1720	−1.7603	0.08093	0.26262
8	0.18160	0.81840	−0.8936	−0.2004	−1.6074	0.09847	0.29037
9	0.20519	0.79481	−0.8110	−0.2297	−1.4712	0.11660	0.31752
10	0.22877	0.77123	−0.7333	−0.2598	−1.3479	0.13525	0.34415
11	0.25236	0.74764	−0.6596	−0.2908	−1.2350	0.15436	0.37032
12	0.27594	0.72406	−0.5891	−0.3229	−1.1305	0.17389	0.39607
13	0.29953	0.70047	−0.5212	−0.3560	−1.0328	0.19379	0.42143
14	0.32311	0.67689	−0.4555	−0.3903	−0.9410	0.21406	0.44644
15	0.34670	0.65330	−0.3916	−0.4257	−0.8540	0.23466	0.47110
16	0.37028	0.62972	−0.3291	−0.4625	−0.7711	0.25557	0.49546
17	0.39387	0.60613	−0.2678	−0.5007	−0.6918	0.27679	0.51950
18	0.41745	0.58255	−0.2074	−0.5403	−0.6155	0.29831	0.54326
19	0.44104	0.55896	−0.1477	−0.5817	−0.5418	0.32011	0.56672
20	0.46462	0.53538	−0.0885	−0.6248	−0.4704	0.34219	0.58991
21	0.48821	0.51179	−0.0295	−0.6698	−0.4007	0.36455	0.61281

Table T2 *Cont.*

i	F'(i)	1 − F'(i)	Xn(i)	Xe(i)	Xw(i)	p0.05	p0.95
22	0.51179	0.48821	0.0295	−0.7170	−0.3327	0.38719	0.63545
23	0.53538	0.46462	0.0886	−0.7665	−0.2659	0.41009	0.65781
24	0.55896	0.44104	0.1480	−0.8186	−0.2001	0.43328	0.67989
25	0.58255	0.41745	0.2080	−0.8736	−0.1352	0.45674	0.70169
26	0.60613	0.39387	0.2688	−0.9317	−0.0707	0.48050	0.72320
27	0.62972	0.37028	0.3307	−0.9935	−0.0065	0.50454	0.74443
28	0.65330	0.34670	0.3938	−1.0593	0.0576	0.52889	0.76534
29	0.67689	0.32311	0.4586	−1.1298	0.1220	0.55356	0.78594
30	0.70047	0.29953	0.5254	−1.2055	0.1869	0.57857	0.80621
31	0.72406	0.27594	0.5946	−1.2876	0.2527	0.60393	0.82611
32	0.74764	0.25236	0.6668	−1.3769	0.3198	0.62968	0.84564
33	0.77123	0.22877	0.7426	−1.4750	0.3887	0.56585	0.86475
34	0.79481	0.20519	0.8230	−1.5838	0.4598	0.68248	0.88340
35	0.81840	0.18160	0.9092	−1.7059	0.5341	0.70963	0.90153
36	0.84198	0.15802	1.0026	−1.8450	0.6125	0.73738	0.91907
37	0.86557	0.13443	1.1057	−2.0067	0.6965	0.76584	0.93591
38	0.88915	0.11085	1.2222	−2.1996	0.7883	0.79517	0.95190
39	0.91274	0.08726	1.3580	−2.4388	0.8915	0.82561	0.96681
40	0.93632	0.06368	1.5249	−2.7539	1.0130	0.85759	0.98025
41	0.95991	0.04009	1.7500	−3.2165	1.1683	0.89196	0.99147
42	0.98349	0.01651	2.1323	−4.1038	1.4119	0.93116	0.99878
			S3: 0.5870	−40.9930	−23.4619		
			S4: 36.6518	75.2004	73.5021		

Total number of results: 43

i	F'(i)	1 − F'(i)	Xn(i)	Xe(i)	Xw(i)	p0.05	p0.95
1	0.01613	0.98387	−1.9237	−0.0163	−4.1190	0.00119	0.06730
2	0.03917	0.96083	−1.6489	−0.0400	−3.2199	0.00833	0.10563
3	0.06221	0.93779	−1.4652	−0.0642	−2.7453	0.01928	0.13927
4	0.08525	0.91475	−1.3209	−0.0891	−2.4179	0.03240	0.17056
5	0.10829	0.89171	−1.1993	−0.1146	−2.1661	0.04695	0.20036
6	0.13134	0.86866	−1.0927	−0.1408	−1.9604	0.06256	0.22907
7	0.15438	0.84562	−0.9968	−0.1677	−1.7857	0.07898	0.25694
8	0.17742	0.82258	−0.9089	−0.1953	−1.6332	0.09609	0.28413
9	0.20046	0.79954	−0.8272	−0.2237	−1.4974	0.11377	0.31073
10	0.22350	0.77650	−0.7503	−0.2530	−1.3745	0.13195	0.33682
11	0.24654	0.75346	−0.6775	−0.2831	−1.2620	0.15058	0.36247
12	0.26959	0.73041	−0.6078	−0.3141	−1.1579	0.16961	0.38772
13	0.29263	0.70737	−0.5408	−0.3462	−1.0607	0.18901	0.41259
14	0.31567	0.68433	−0.4760	−0.3793	−0.9694	0.20875	0.43712
15	0.33871	0.66129	−0.4131	−0.4136	−0.8829	0.22881	0.46132
16	0.36175	0.63825	−0.3516	−0.4490	−0.8007	0.24918	0.48522
17	0.38479	0.61521	−0.2913	−0.4858	−0.7220	0.26984	0.50883
18	0.40783	0.59217	−0.2320	−0.5240	−0.6463	0.29078	0.53215
19	0.43088	0.56912	−0.1734	−0.5637	−0.5733	0.31200	0.55520
20	0.45392	0.54608	−0.1153	−0.6050	−0.5026	0.33348	0.57799
21	0.47696	0.52304	−0.0576	−0.6481	−0.4337	0.35522	0.60051
22	0.50000	0.50000	0.0000	−0.6931	−0.3665	0.37722	0.62273
23	0.52304	0.47696	0.0576	−0.7403	−0.3007	0.39949	0.64478
24	0.54608	0.45392	0.1155	−0.7898	−0.2359	0.42201	0.66652

Table T2 *Cont.*

i	$F'(i)$	$1 - F'(i)$	Xn(i)	Xe(i)	Xw(i)	p0.05	p0.95
25	0.56912	0.43088	0.1738	−0.8419	−0.1721	0.44480	0.68800
26	0.59217	0.40783	0.2327	−0.8969	−0.1088	0.46785	0.70922
27	0.61521	0.38479	0.2925	−0.9551	−0.0460	0.49117	0.73016
28	0.63825	0.36175	0.3533	−1.0168	0.0167	0.51478	0.75082
29	0.66129	0.33871	0.4155	−1.0826	0.0794	0.53868	0.77119
30	0.68433	0.31567	0.4794	−1.1531	0.1424	0.56288	0.79125
31	0.70737	0.29263	0.5453	−1.2289	0.2061	0.58741	0.81099
32	0.73041	0.26959	0.6137	−1.3109	0.2707	0.61228	0.83039
33	0.75346	0.24654	0.6851	−1.4002	0.3366	0.63753	0.84942
34	0.77650	0.22350	0.7602	−1.4983	0.4044	0.66318	0.86805
35	0.79954	0.20046	0.8398	−1.6071	0.4745	0.68927	0.88623
36	0.82258	0.17742	0.9252	−1.7292	0.5477	0.71587	0.90391
37	0.84562	0.15438	1.0178	−1.8684	0.6251	0.74306	0.92102
38	0.86866	0.13134	1.1202	−2.0300	0.7080	0.77093	0.93744
39	0.89171	0.10829	1.2358	−2.2229	0.7988	0.79964	0.95305
40	0.91475	0.08525	1.3708	−2.4621	0.9010	0.82944	0.96760
41	0.93779	0.06221	1.5368	−2.7772	1.0214	0.86073	0.98071
42	0.96083	0.03917	1.7608	−3.2398	1.1755	0.89437	0.99167
43	0.98387	0.01613	2.1416	−4.1271	1.4176	0.93270	0.99881
			S3: 0.6033	−41.9883	−24.0351		
			S4: 37.5795	77.1347	75.4119		

Total number of results: 44

i	$F'(i)$	$1 - F'(i)$	Xn(i)	Xe(i)	Xw(i)	p0.05	p0.95
1	0.01577	0.98423	−1.9296	−0.0159	−4.1420	0.00116	0.06582
2	0.03829	0.96171	−1.6572	−0.0390	−3.2432	0.00814	0.10334
3	0.06081	0.93919	−1.4750	−0.0627	−2.7688	0.01884	0.13626
4	0.08333	0.91667	−1.3319	−0.0870	−2.4417	0.03165	0.16690
5	0.10586	0.89414	−1.2113	−0.1119	−2.1903	0.04586	0.19608
6	0.12838	0.87162	−1.1057	−0.1374	−1.9849	0.06109	0.22420
7	0.15090	0.84910	−1.0107	−0.1636	−1.8105	0.07713	0.25151
8	0.17342	0.82658	−0.9236	−0.1905	−1.6583	0.09382	0.27814
9	0.19595	0.80405	−0.8427	−0.2181	−1.5229	0.11107	0.30422
10	0.21847	0.78153	−0.7667	−0.2465	−1.4004	0.12881	0.32980
11	0.24099	0.75901	−0.6947	−0.2757	−1.2883	0.14698	0.35495
12	0.26351	0.73649	−0.6259	−0.3059	−1.1846	0.16554	0.37971
13	0.28604	0.71396	−0.5597	−0.3369	−1.0879	0.18445	0.40410
14	0.30856	0.69144	−0.4958	−0.3690	−0.9970	0.20370	0.42817
15	0.33108	0.66892	−0.4337	−0.4021	−0.9111	0.22326	0.45193
16	0.35360	0.64640	−0.3732	−0.4363	−0.8293	0.24311	0.47539
17	0.37613	0.62387	−0.3139	−0.4718	−0.7512	0.26323	0.49857
18	0.39865	0.60135	−0.2555	−0.5086	−0.6761	0.28363	0.52148
19	0.42117	0.57883	−0.1980	−0.5467	−0.6038	0.30429	0.54413
20	0.44369	0.55631	−0.1410	−0.5864	−0.5337	0.32520	0.56653
21	0.46622	0.53378	−0.0845	−0.6278	−0.4656	0.34636	0.58867
22	0.48874	0.51126	−0.0281	−0.6709	−0.3992	0.36777	0.61057
23	0.51126	0.48874	0.0282	−0.7159	−0.3342	0.38943	0.63223
24	0.53378	0.46622	0.0846	−0.7631	−0.2704	0.41133	0.65363
25	0.55631	0.44369	0.1413	−0.8126	−0.2075	0.43347	0.67480
26	0.57883	0.42117	0.1985	−0.8647	−0.1454	0.45587	0.69571

Table T2 *Cont.*

i	F'(i)	1 − F'(i)	Xn(i)	Xe(i)	Xw(i)	p0.05	p0.95
27	0.60135	0.39865	0.2564	−0.9197	−0.0837	0.47852	0.71637
28	0.62387	0.37613	0.3152	−0.9778	−0.0224	0.50143	0.73677
29	0.64640	0.35360	0.3752	−1.0396	0.0388	0.52461	0.75689
30	0.66892	0.33108	0.4365	−1.1054	0.1002	0.54807	0.77674
31	0.69144	0.30856	0.4995	−1.1758	0.1620	0.57183	0.79630
32	0.71396	0.28604	0.5646	−1.2516	0.2245	0.59590	0.81554
33	0.73649	0.26351	0.6322	−1.3337	0.2879	0.62029	0.83446
34	0.75901	0.24099	0.7028	−1.4230	0.3528	0.64505	0.85302
35	0.78153	0.21847	0.7772	−1.5211	0.4194	0.67020	0.87119
36	0.80405	0.19595	0.8560	−1.6299	0.4885	0.69578	0.88892
37	0.82658	0.17342	0.9406	−1.7520	0.5608	0.72185	0.90618
38	0.84910	0.15090	1.0326	−1.8911	0.6372	0.74849	0.92287
39	0.87162	0.12838	1.1342	−2.0528	0.7192	0.77580	0.93891
40	0.89414	0.10586	1.2490	−2.2457	0.8090	0.80392	0.95414
41	0.91667	0.08333	1.3832	−2.4849	0.9102	0.83310	0.96835
42	0.93919	0.06081	1.5483	−2.8000	1.0296	0.86374	0.98116
43	0.96171	0.03829	1.7713	−3.2626	1.1825	0.89666	0.99186
44	0.98423	0.01577	2.1507	−4.1499	1.4231	0.93418	0.99883
			S3: 0.6196	−42.9838	−24.6083		
			S4: 38.5073	79.0703	77.3231		

Total number of results: 45

i	F'(i)	1 − F'(i)	Xn(i)	Xe(i)	Xw(i)	p0.05	p0.95
1	0.01542	0.98458	−1.9353	−0.0155	−4.1644	0.00114	0.06440
2	0.03744	0.96256	−1.6652	−0.0382	−3.2659	0.00795	0.10113
3	0.05947	0.94053	−1.4845	−0.0613	−2.7918	0.01842	0.13338
4	0.08150	0.91850	−1.3426	−0.0850	−2.4650	0.03093	0.16339
5	0.10352	0.89648	−1.2230	−0.1093	−2.2138	0.04481	0.19198
6	0.12555	0.87445	−1.1183	−0.1342	−2.0087	0.05969	0.21954
7	0.14758	0.85242	−1.0242	−0.1597	−1.8346	0.07536	0.24630
8	0.16960	0.83040	−0.9379	−0.1859	−1.6828	0.09166	0.27241
9	0.19163	0.80837	−0.8578	−0.2127	−1.5477	0.10850	0.29797
10	0.21366	0.78634	−0.7826	−0.2404	−1.4256	0.12582	0.32306
11	0.23568	0.76432	−0.7114	−0.2688	−1.3139	0.14355	0.34773
12	0.25771	0.74229	−0.6433	−0.2980	−1.2106	0.16166	0.37202
13	0.27974	0.72026	−0.5780	−0.3281	−1.1143	0.18012	0.39596
14	0.30176	0.69824	−0.5149	−0.3592	−1.0239	0.19889	0.41958
15	0.32379	0.67621	−0.4537	−0.3912	−0.9384	0.21796	0.44290
16	0.34582	0.65419	−0.3940	−0.4244	−0.8572	0.23732	0.46594
17	0.36784	0.63216	−0.3355	−0.4586	−0.7795	0.25694	0.48871
18	0.38987	0.61013	−0.2782	−0.4941	−0.7051	0.27683	0.51122
19	0.41189	0.58811	−0.2216	−0.5308	−0.6333	0.29696	0.53348
20	0.43392	0.56608	−0.1657	−0.5690	−0.5638	0.31733	0.55549
21	0.45595	0.54405	−0.1102	−0.6087	−0.4964	0.33794	0.57727
22	0.47797	0.52203	−0.0551	−0.6500	−0.4307	0.35879	0.59882
23	0.50000	0.50000	0.0000	−0.6931	−0.3665	0.37987	0.62013
24	0.52203	0.47797	0.0551	−0.7382	−0.3035	0.40118	0.64121
25	0.54405	0.45595	0.1104	−0.7854	−0.2416	0.42273	0.66205
26	0.56608	0.43392	0.1661	−0.8349	−0.1805	0.44451	0.68267
27	0.58811	0.41189	0.2223	−0.8870	−0.1199	0.46652	0.70304

Table T2 *Cont.*

i	$F'(i)$	$1 - F'(i)$	Xn(i)	Xe(i)	Xw(i)	p0.05	p0.95
28	0.61013	0.38987	0.2792	−0.9419	−0.0598	0.48878	0.72317
29	0.63216	0.36784	0.3371	−1.0001	0.0001	0.51129	0.74306
30	0.65419	0.34582	0.3962	−1.0619	0.0600	0.53406	0.76368
31	0.67621	0.32379	0.4567	−1.1277	0.1201	0.55710	0.78203
32	0.69824	0.30176	0.5189	−1.1981	0.1808	0.58042	0.80111
33	0.72026	0.27974	0.5833	−1.2739	0.2421	0.60404	0.81988
34	0.74229	0.25771	0.6501	−1.3559	0.3045	0.62798	0.83834
35	0.76432	0.23568	0.7200	−1.4453	0.3683	0.65227	0.85645
36	0.78634	0.21366	0.7936	−1.5434	0.4340	0.67694	0.87418
37	0.80837	0.19163	0.8718	−1.6522	0.5021	0.70203	0.89150
38	0.83040	0.16960	0.9557	−1.7743	0.5734	0.72759	0.90834
39	0.85242	0.14758	1.0469	−1.9134	0.6489	0.75370	0.92464
40	0.87445	0.12555	1.1478	−2.0750	0.7300	0.78046	0.94030
41	0.89648	0.10352	1.2619	−2.2679	0.8189	0.80802	0.95518
42	0.91850	0.08150	1.3953	−2.5072	0.9192	0.83661	0.96907
43	0.94053	0.05947	1.5596	−2.8223	1.0375	0.86662	0.98158
44	0.96256	0.03744	1.7815	−3.2849	1.1893	0.89887	0.99205
45	0.98458	0.01542	2.1596	−4.1722	1.4284	0.93560	0.99886
			S3: 0.6359	−43.9793	−25.1817		
			S4: 39.4355	81.0071	79.2355		

Total number of results: 46

i	$F'(i)$	$1 - F'(i)$	Xn(i)	Xe(i)	Xw(i)	p0.05	p0.95
1	0.01509	0.98491	−1.9408	−0.0152	−4.1864	0.00111	0.06305
2	0.03664	0.96336	−1.6730	−0.0373	−3.2881	0.00778	0.09902
3	0.05819	0.94181	−1.4937	−0.0600	−2.8142	0.01801	0.13061
4	0.07974	0.92026	−1.3529	−0.0831	−2.4877	0.03025	0.16002
5	0.10129	0.89871	−1.2344	−0.1068	−2.2368	0.04382	0.18804
6	0.12284	0.87716	−1.1306	−0.1311	−2.0320	0.05836	0.21506
7	0.14440	0.85560	−1.0372	−0.1559	−1.8582	0.07366	0.24130
8	0.16595	0.83405	−0.9518	−0.1815	−1.7067	0.08959	0.26691
9	0.18750	0.81250	−0.8724	−0.2076	−1.5720	0.10605	0.29198
10	0.20905	0.79095	−0.7980	−0.2345	−1.4502	0.12296	0.31659
11	0.23060	0.76940	−0.7275	−0.2621	−1.3388	0.14028	0.34079
12	0.25216	0.74784	−0.6602	−0.2906	−1.2359	0.15796	0.36463
13	0.27371	0.72629	−0.5956	−0.3198	−1.1401	0.17598	0.38813
14	0.29526	0.70474	−0.5333	−0.3499	−1.0500	0.19430	0.41132
15	0.31681	0.68319	−0.4729	−0.3810	−0.9650	0.21292	0.43422
16	0.33836	0.66164	−0.4140	−0.4130	−0.8842	0.23180	0.45665
17	0.35991	0.64009	−0.3564	−0.4462	−0.8071	0.25085	0.47922
18	0.38147	0.61853	−0.2999	−0.4804	−0.7331	0.27034	0.50134
19	0.40302	0.59698	−0.2443	−0.5159	−0.6619	0.28997	0.52322
20	0.42457	0.57543	−0.1894	−0.5526	−0.5931	0.30984	0.54487
21	0.44612	0.55388	−0.1349	−0.5908	−0.5263	0.32993	0.56629
22	0.46767	0.53233	−0.0808	−0.6305	−0.4612	0.35025	0.58749
23	0.48922	0.51078	−0.0269	−0.6718	−0.3978	0.37078	0.60846
24	0.51078	0.48922	0.0269	−0.7149	−0.3356	0.39154	0.62922
25	0.53233	0.46767	0.0809	−0.7600	−0.2745	0.41251	0.64975
26	0.55388	0.44612	0.1352	−0.8072	−0.2142	0.43371	0.67007
27	0.57543	0.42457	0.1899	−0.8567	−0.1547	0.45513	0.69016

Table T2 *Cont.*

i	F'(i)	1 − F'(i)	Xn(i)	Xe(i)	Xw(i)	p0.05	p0.95
28	0.59698	0.40302	0.2451	−0.9088	−0.0957	0.47678	0.71002
29	0.61853	0.38147	0.3012	−0.9637	−0.0369	0.49866	0.72966
30	0.64009	0.35991	0.3582	−1.0219	0.0217	0.52078	0.74905
31	0.66164	0.33836	0.4165	−1.0836	0.0803	0.54315	0.76819
32	0.68319	0.31681	0.4762	−1.1495	0.1393	0.56578	0.78708
33	0.70474	0.29526	0.5377	−1.2199	0.1988	0.58868	0.80569
34	0.72629	0.27371	0.6013	−1.2957	0.2590	0.61187	0.82402
35	0.74784	0.25216	0.6674	−1.3777	0.3204	0.63537	0.84204
36	0.76940	0.23060	0.7366	−1.4671	0.3833	0.65921	0.85972
37	0.79095	0.20905	0.8095	−1.5652	0.4480	0.68341	0.87704
38	0.81250	0.18750	0.8870	−1.6740	0.5152	0.70805	0.89395
39	0.83405	0.16595	0.9702	−1.7961	0.5856	0.73309	0.91041
40	0.85560	0.14440	1.0608	−1.9352	0.6602	0.75870	0.92633
41	0.87716	0.12284	1.1610	−2.0968	0.7404	0.78494	0.94164
42	0.89871	0.10129	1.2744	−2.2897	0.8284	0.81196	0.95618
43	0.92026	0.07974	1.4071	−2.5290	0.9278	0.83998	0.96975
44	0.94181	0.05819	1.5705	−2.8440	1.0452	0.86939	0.98199
45	0.96336	0.03664	1.7915	−3.3067	1.1959	0.90098	0.99222
46	0.98491	0.01509	2.1683	−4.1940	1.4336	0.93695	0.99889
S3:			0.6522	−44.9749	−25.7552		
S4:			40.3638	82.9450	81.1492		

Total number of results: 47

i	F'(i)	1 − F'(i)	Xn(i)	Xe(i)	Xw(i)	p0.05	p0.95
1	0.01477	0.98523	−1.9462	−0.0149	−4.2079	0.00109	0.06175
2	0.03586	0.96414	−1.6805	−0.0365	−3.3098	0.00761	0.09700
3	0.05696	0.94304	−1.5027	−0.0586	−2.8362	0.01762	0.12796
4	0.07806	0.92194	−1.3630	−0.0813	−2.5099	0.02959	0.15679
5	0.09916	0.90084	−1.2454	−0.1044	−2.2593	0.04286	0.18427
6	0.12025	0.87975	−1.1424	−0.1281	−2.0548	0.05708	0.21076
7	0.14135	0.85865	−1.0499	−0.1524	−1.8813	0.07205	0.23650
8	0.16245	0.83755	−0.9652	−0.1773	−1.7301	0.08762	0.26162
9	0.18354	0.81646	−0.8866	−0.2028	−1.5956	0.10370	0.28622
10	0.20464	0.79536	−0.8129	−0.2290	−1.4742	0.12023	0.31037
11	0.22574	0.77426	−0.7431	−0.2558	−1.3632	0.13715	0.33413
12	0.24684	0.75316	−0.6765	−0.2835	−1.2606	0.15443	0.35753
13	0.26793	0.73207	−0.6127	−0.3119	−1.1651	0.17203	0.38060
14	0.28903	0.71097	−0.5511	−0.3411	−1.0755	0.18993	0.40338
15	0.31013	0.68987	−0.4914	−0.3712	−0.9909	0.20810	0.42587
16	0.33122	0.66878	−0.4334	−0.4023	−0.9105	0.22654	0.44810
17	0.35232	0.64768	−0.3766	−0.4344	−0.8339	0.24523	0.47009
18	0.37342	0.62658	−0.3209	−0.4675	−0.7604	0.26416	0.49183
19	0.39451	0.60549	−0.2662	−0.5017	−0.6897	0.28331	0.51334
20	0.41561	0.58439	−0.2121	−0.5372	−0.6214	0.30269	0.53463
21	0.43671	0.56329	−0.1587	−0.5740	−0.5552	0.32229	0.55570
22	0.45781	0.54219	−0.1056	−0.6121	−0.4908	0.34210	0.57556
23	0.47890	0.52110	−0.0527	−0.6518	−0.4280	0.36212	0.59721
24	0.50000	0.50000	0.0000	−0.6931	−0.3665	0.38235	0.61765
25	0.52110	0.47890	0.0528	−0.7363	−0.3062	0.40279	0.63787
26	0.54219	0.45781	0.1057	−0.7813	−0.2468	0.42344	0.65790

Table T2 *Cont.*

i	F'(i)	1 − F'(i)	Xn(i)	Xe(i)	Xw(i)	p0.05	p0.95
27	0.56329	0.43671	0.1590	−0.8285	−0.1882	0.44430	0.67771
28	0.58439	0.41561	0.2127	−0.8780	−0.1301	0.46537	0.69730
29	0.60549	0.39451	0.2671	−0.9301	−0.0725	0.48666	0.71668
30	0.62658	0.37342	0.3224	−0.9851	−0.0151	0.50817	0.73584
31	0.64768	0.35232	0.3786	−1.0432	0.0423	0.52991	0.75477
32	0.66878	0.33122	0.4361	−1.1050	0.0998	0.55190	0.77346
33	0.68987	0.31013	0.4951	−1.1708	0.1577	0.57413	0.79190
34	0.71097	0.28903	0.5558	−1.2412	0.2161	0.59662	0.81007
35	0.73207	0.26793	0.6187	−1.3170	0.2754	0.61940	0.82797
36	0.75316	0.24684	0.6842	−1.3990	0.3358	0.64247	0.84557
37	0.77426	0.22574	0.7527	−1.4884	0.3977	0.66587	0.86285
38	0.79536	0.20464	0.8250	−1.5865	0.4615	0.68963	0.87977
39	0.81646	0.18354	0.9018	−1.6953	0.5279	0.71378	0.89630
40	0.83755	0.16245	0.9844	−1.8174	0.5974	0.73838	0.91238
41	0.85865	0.14135	1.0743	−1.9565	0.6712	0.76350	0.92795
42	0.87975	0.12025	1.1738	−2.1182	0.7505	0.78924	0.94291
43	0.90084	0.09916	1.2866	−2.3111	0.8377	0.81573	0.95714
44	0.92194	0.07806	1.4185	−2.5503	0.9362	0.84321	0.97041
45	0.94304	0.05696	1.5811	−2.8654	1.0527	0.87204	0.98238
46	0.96414	0.03586	1.8012	−3.3280	1.2024	0.90300	0.99239
47	0.98523	0.01477	2.1767	−4.2153	1.4387	0.93825	0.99891
			S3: 0.6686	−45.9707	−26.3287		
			S4: 41.2923	84.8842	83.0642		

Total number of results: 48

i	F'(i)	1 − F'(i)	Xn(i)	Xe(i)	Xw(i)	p0.05	p0.95
1	0.01446	0.98554	−1.9514	−0.0146	−4.2289	0.00107	0.06050
2	0.03512	0.96488	−1.6879	−0.0358	−3.3311	0.00745	0.09506
3	0.05579	0.94421	−1.5114	−0.0574	−2.8577	0.01725	0.12541
4	0.07645	0.92355	−1.3728	−0.0795	−2.5317	0.02897	0.15369
5	0.09711	0.90289	−1.2561	−0.1022	−2.2813	0.04195	0.18064
6	0.11777	0.88223	−1.1539	−0.1253	−2.0770	0.05586	0.20663
7	0.13843	0.86157	−1.0622	−0.1490	−1.9038	0.07050	0.23188
8	0.15909	0.84091	−0.9782	−0.1733	−1.7529	0.08573	0.25623
9	0.17975	0.82025	−0.9004	−0.1981	−1.6187	0.10146	0.28068
10	0.20041	0.79959	−0.8273	−0.2237	−1.4976	0.11762	0.30439
11	0.22107	0.77893	−0.7582	−0.2498	−1.3869	0.13416	0.32772
12	0.24174	0.75826	−0.6924	−0.2767	−1.2847	0.15105	0.35069
13	0.26240	0.73760	−0.6292	−0.3043	−1.1896	0.16825	0.37336
14	0.28306	0.71694	−0.5684	−0.3328	−1.1003	0.18574	0.39573
15	0.30372	0.69628	−0.5094	−0.3620	−1.0161	0.20350	0.41783
16	0.32438	0.67562	−0.4520	−0.3921	−0.9362	0.22151	0.43968
17	0.34504	0.65496	−0.3960	−0.4232	−0.8600	0.23977	0.46129
18	0.36570	0.63430	−0.3412	−0.4552	−0.7869	0.25825	0.48266
19	0.38636	0.61364	−0.2872	−0.4884	−0.7167	0.27696	0.50382
20	0.40702	0.59298	−0.2341	−0.5226	−0.6489	0.29588	0.52476
21	0.42769	0.57231	−0.1815	−0.5581	−0.5833	0.31500	0.54549
22	0.44835	0.55165	−0.1293	−0.5948	−0.5195	0.33434	0.56602
23	0.46901	0.53099	−0.0775	−0.6330	−0.4573	0.35387	0.58634
24	0.48967	0.51033	−0.0258	−0.6727	−0.3965	0.37360	0.60647

Table T2 *Cont.*

i	F'(i)	1 − F'(i)	Xn(i)	Xe(i)	Xw(i)	p0.05	p0.95
25	0.51033	0.48967	0.0258	−0.7140	−0.3368	0.39353	0.62640
26	0.53099	0.46901	0.0776	−0.7571	−0.2782	0.41366	0.64613
27	0.55165	0.44835	0.1296	−0.8022	−0.2204	0.43398	0.66566
28	0.57231	0.42769	0.1819	−0.8494	−0.1633	0.45451	0.68500
29	0.59298	0.40702	0.2348	−0.8989	−0.1066	0.47524	0.70412
30	0.61364	0.38636	0.2884	−0.9510	−0.0503	0.49618	0.72304
31	0.63430	0.36570	0.3428	−1.0059	0.0059	0.51734	0.74175
32	0.65496	0.34504	0.3983	−1.0641	0.0621	0.53871	0.76023
33	0.67562	0.32438	0.4551	−1.1258	0.1185	0.56032	0.77848
34	0.69628	0.30372	0.5133	−1.1917	0.1753	0.58217	0.79650
35	0.71694	0.28306	0.5734	−1.2621	0.2328	0.60427	0.81426
36	0.73760	0.26240	0.6356	−1.3379	0.2911	0.62664	0.83175
37	0.75826	0.24174	0.7005	−1.4199	0.3506	0.64931	0.84895
38	0.77893	0.22107	0.7683	−1.5093	0.4116	0.67228	0.86584
39	0.79959	0.20041	0.8400	−1.6074	0.4746	0.69561	0.88238
40	0.82025	0.17975	0.9162	−1.7162	0.5401	0.71932	0.89854
41	0.84091	0.15909	0.9982	−1.8383	0.6088	0.74347	0.91427
42	0.86157	0.13843	1.0874	−1.9774	0.6818	0.76812	0.92950
43	0.88223	0.11777	1.1863	−2.1390	0.7604	0.79337	0.94414
44	0.90289	0.09711	1.2984	−2.3319	0.8467	0.81926	0.95805
45	0.92355	0.07645	1.4297	−2.5712	0.9444	0.84631	0.97103
46	0.94421	0.05579	1.5915	−2.8862	1.0600	0.87459	0.98275
47	0.96488	0.03512	1.8107	−3.3489	1.2086	0.90494	0.99255
48	0.98554	0.01446	2.1849	−4.2362	1.4437	0.93950	0.99893
			S3: 0.6850	−46.9665	−26.9023		
			S4: 42.2211	86.8244	84.9802		

Total number of results: 49

i	F'(i)	1 − F'(i)	Xn(i)	Xe(i)	Xw(i)	p0.05	p0.95
1	0.01417	0.98583	−1.9565	−0.0143	−4.2495	0.00105	0.05931
2	0.03441	0.96559	−1.6950	−0.0350	−3.3519	0.00730	0.09319
3	0.05466	0.94534	−1.5198	−0.0562	−2.8787	0.01689	0.12297
4	0.07490	0.92510	−1.3823	−0.0779	−2.5529	0.02836	0.15071
5	0.09514	0.90486	−1.2665	−0.1000	−2.3028	0.04108	0.17715
6	0.11538	0.88462	−1.1651	−0.1226	−2.0988	0.05469	0.20266
7	0.13563	0.86437	−1.0741	−0.1458	−1.9259	0.06902	0.22744
8	0.15587	0.84413	−0.9909	−0.1694	−1.7752	0.08392	0.25164
9	0.17611	0.82389	−0.9137	−0.1937	−1.6413	0.09931	0.27535
10	0.19636	0.80364	−0.8413	−0.2186	−1.5205	0.11512	0.29864
11	0.21660	0.78340	−0.7729	−0.2441	−1.4101	0.13130	0.32154
12	0.23684	0.76316	−0.7077	−0.2703	−1.3083	0.14782	0.34411
13	0.25709	0.74292	−0.6452	−0.2972	−1.2134	0.16464	0.36698
14	0.27733	0.72267	−0.5850	−0.3248	−1.1245	0.18174	0.38836
15	0.29757	0.70243	−0.5268	−0.3532	−1.0407	0.19910	0.41008
16	0.31781	0.68219	−0.4701	−0.3825	−0.9612	0.21671	0.43156
17	0.33806	0.66194	−0.4148	−0.4126	−0.8853	0.23455	0.45280
18	0.35830	0.64170	−0.3607	−0.4436	−0.8128	0.25261	0.47382
19	0.37854	0.62146	−0.3076	−0.4757	−0.7430	0.27088	0.49463
20	0.39879	0.60121	−0.2552	−0.5088	−0.6757	0.28936	0.51523
21	0.41903	0.58097	−0.2034	−0.5431	−0.6105	0.30804	0.53563

Table T2 *Cont.*

i	F'(i)	1−F'(i)	Xn(i)	Xe(i)	Xw(i)	p0.05	p0.95
22	0.43927	0.56073	−0.1522	−0.5785	−0.5473	0.32692	0.55584
23	0.45951	0.54049	−0.1013	−0.6153	−0.4857	0.34599	0.57585
24	0.47976	0.52024	−0.0506	−0.6535	−0.4255	0.36524	0.59568
25	0.50000	0.50000	0.0000	−0.6931	−0.3665	0.38469	0.61531
26	0.52024	0.47976	0.0506	−0.7345	−0.3086	0.40432	0.63476
27	0.54049	0.45951	0.1014	−0.7776	−0.2516	0.42415	0.65401
28	0.56073	0.43927	0.1525	−0.8226	−0.1952	0.44416	0.67308
29	0.58097	0.41903	0.2040	−0.8698	−0.1395	0.46436	0.69196
30	0.60121	0.39879	0.2561	−0.9193	−0.0841	0.48477	0.71064
31	0.62146	0.37854	0.3089	−0.9714	−0.0290	0.50537	0.72912
32	0.64170	0.35830	0.3626	−1.0264	0.0260	0.52617	0.74739
33	0.66194	0.33806	0.4173	−1.0845	0.0812	0.54720	0.76545
34	0.68219	0.31781	0.4734	−1.1463	0.1365	0.56844	0.78329
35	0.70243	0.29757	0.5310	−1.2121	0.1924	0.58991	0.80090
36	0.72267	0.27733	0.5904	−1.2826	0.2489	0.61164	0.81826
37	0.74292	0.25709	0.6520	−1.3583	0.3063	0.63362	0.83536
38	0.76316	0.23684	0.7162	−1.4404	0.3649	0.65589	0.85218
39	0.78340	0.21660	0.7835	−1.5297	0.4251	0.67846	0.86870
40	0.80364	0.19636	0.8546	−1.6278	0.4872	0.70136	0.88488
41	0.82389	0.17611	0.9302	−1.7366	0.5519	0.72465	0.90069
42	0.84413	0.15587	1.0116	−1.8587	0.6199	0.74836	0.91608
43	0.86437	0.13563	1.1002	−1.9978	0.6921	0.77256	0.93098
44	0.88462	0.11538	1.1985	−2.1595	0.7699	0.79734	0.94531
45	0.90486	0.09514	1.3099	−2.3524	0.8554	0.82285	0.95892
46	0.92510	0.07490	1.4405	−2.5916	0.9523	0.84929	0.97163
47	0.94534	0.05466	1.6016	−2.9067	1.0670	0.87703	0.98311
48	0.96559	0.03441	1.8200	−3.3693	1.2147	0.90681	0.99270
49	0.98583	0.01417	2.1930	−4.2566	1.4485	0.94069	0.99895
			S3: 0.7013	−47.9624	−27.4759		
			S4: 43.1501	87.7656	86.8974		

Total number of results: 50

i	F'(i)	1−F'(i)	Xn(i)	Xe(i)	Xw(i)	p0.05	p0.95
1	0.01389	0.98611	−1.9614	−0.0140	−4.2697	0.00102	0.05816
2	0.03373	0.96627	−1.7019	−0.0343	−3.3723	0.00715	0.09140
3	0.05357	0.94643	−1.5281	−0.0551	−2.8993	0.01655	0.12061
4	0.07341	0.92659	−1.3915	−0.0762	−2.5738	0.02779	0.14784
5	0.09325	0.90675	−1.2766	−0.0979	−2.3239	0.04024	0.17379
6	0.11310	0.88690	−1.1760	−0.1200	−2.1201	0.05357	0.19883
7	0.13294	0.86706	−1.0857	−0.1426	−1.9474	0.06760	0.22317
8	0.15278	0.84722	−1.0032	−0.1658	−1.7970	0.08218	0.24694
9	0.17262	0.82738	−0.9266	−0.1895	−1.6634	0.09725	0.27022
10	0.19246	0.80754	−0.8549	−0.2138	−1.5429	0.11272	0.29309
11	0.21230	0.78770	−0.7871	−0.2386	−1.4328	0.12856	0.31560
12	0.23214	0.76786	−0.7226	−0.2642	−1.3312	0.14472	0.33778
13	0.25198	0.74802	−0.6607	−0.2903	−1.2367	0.16117	0.35933
14	0.27183	0.72817	−0.6012	−0.3172	−1.1482	0.17790	0.38126
15	0.29167	0.70833	−0.5436	−0.3448	−1.0647	0.19488	0.40262
16	0.31151	0.68849	−0.4876	−0.3733	−0.9855	0.21210	0.42373
17	0.33135	0.66865	−0.4330	−0.4025	−0.9101	0.22955	0.44462

Table T2 *Cont.*

i	$F'(i)$	$1 - F'(i)$	Xn(i)	Xe(i)	Xw(i)	p0.05	p0.95
18	0.35119	0.64881	−0.3796	−0.4326	−0.8379	0.24721	0.46530
19	0.37103	0.62897	−0.3272	−0.4637	−0.7686	0.26507	0.48577
20	0.39087	0.60913	−0.2756	−0.4957	−0.7017	0.28313	0.50604
21	0.41071	0.58929	−0.2246	−0.5288	−0.6371	0.30138	0.52612
22	0.43056	0.56944	−0.1742	−0.5631	−0.5743	0.31980	0.54801
23	0.45040	0.54960	−0.1242	−0.5986	−0.5132	0.33845	0.56572
24	0.47024	0.52976	−0.0744	−0.6353	−0.4536	0.35726	0.58524
25	0.49008	0.50992	−0.0248	−0.6735	−0.3953	0.37625	0.60459
26	0.50992	0.49008	0.0248	−0.7132	−0.3380	0.39541	0.62375
27	0.52976	0.47024	0.0745	−0.7545	−0.2817	0.41476	0.64274
28	0.54960	0.45040	0.1244	−0.7976	−0.2261	0.43428	0.66155
29	0.56944	0.43056	0.1746	−0.8427	−0.1712	0.45399	0.68017
30	0.58929	0.41071	0.2253	−0.8899	−0.1167	0.47388	0.69862
31	0.60913	0.39087	0.2766	−0.9394	−0.0625	0.49396	0.71687
32	0.62897	0.37103	0.3287	−0.9915	−0.0086	0.51423	0.73493
33	0.64881	0.35119	0.3817	−1.0464	0.0454	0.53470	0.75279
34	0.66865	0.33135	0.4358	−1.1046	0.0995	0.55538	0.77045
35	0.68849	0.31151	0.4912	−1.1663	0.1539	0.57627	0.78790
36	0.70833	0.29167	0.5481	−1.2321	0.2088	0.59738	0.80511
37	0.72817	0.27183	0.6070	−1.3026	0.2644	0.61874	0.82210
38	0.74802	0.25198	0.6680	−1.3784	0.3209	0.64034	0.83882
39	0.76786	0.23214	0.7316	−1.4604	0.3787	0.66222	0.85528
40	0.78770	0.21230	0.7983	−1.5497	0.4381	0.68440	0.87144
41	0.80754	0.19246	0.8687	−1.6479	0.4995	0.70691	0.88728
42	0.82738	0.17262	0.9438	−1.7567	0.5634	0.72978	0.90275
43	0.84722	0.15278	1.0246	−1.8788	0.6306	0.75306	0.91781
44	0.86706	0.13294	1.1127	−2.0179	0.7020	0.77683	0.93240
45	0.88690	0.11310	1.2104	−2.1795	0.7791	0.80117	0.94643
46	0.90675	0.09325	1.3212	−2.3724	0.8639	0.82621	0.95976
47	0.92659	0.07341	1.4511	−2.6117	0.9600	0.85216	0.97221
48	0.94643	0.05357	1.6115	−2.9267	1.0739	0.87939	0.99345
49	0.96627	0.03373	1.8290	−3.3894	1.2206	0.90860	0.99285
50	0.98611	0.01389	2.2009	−4.2767	1.4532	0.94184	0.99897
			S3: 0.7178	−48.9584	−28.0496		
			S4: 44.0792	90.7079	88.8156		

Table T3 Cumulative probability and corresponding linearized values

$F(.)$	X_n	X_e	X_w
0.001	-2.3522	-0.0010	-6.9072
0.002	-2.2881	-0.0020	-6.2136
0.003	-2.2402	-0.0030	-5.8076
0.004	-2.2007	-0.0040	-5.5194
0.005	-2.1664	-0.0050	-5.2958
0.006	-2.1359	-0.0060	-5.1129
0.007	-2.1082	-0.0070	-4.9583
0.008	-2.0827	-0.0080	-4.8243
0.009	-2.0591	-0.0090	-4.7060
0.010	-2.0369	-0.0101	-4.6001
0.011	-2.0160	-0.0111	-4.5043
0.012	-1.9962	-0.0121	-4.4168
0.013	-1.9774	-0.0131	-4.3362
0.014	-1.9595	-0.0141	-4.2616
0.015	-1.9423	-0.0151	-4.1921
0.016	-1.9258	-0.0161	-4.1271
0.017	-1.9099	-0.0171	-4.0659
0.018	-1.8945	-0.0182	-4.0083
0.019	-1.8797	-0.0192	-3.9537
0.020	-1.8653	-0.0202	-3.9019
0.021	-1.8514	-0.0212	-3.8526
0.022	-1.8379	-0.0222	-3.8056
0.023	-1.8248	-0.0233	-3.7606
0.024	-1.8120	-0.0243	-3.7175
0.025	-1.7995	-0.0253	-3.6762
0.026	-1.7873	-0.0263	-3.6365
0.027	-1.7755	-0.0274	-3.5982
0.028	-1.7639	-0.0284	-3.5613
0.029	-1.7525	-0.0294	-3.5257
0.030	-1.7414	-0.0305	-3.4913
0.031	-1.7305	-0.0315	-3.4580
0.032	-1.7199	-0.0325	-3.4258
0.033	-1.7094	-0.0336	-3.3945
0.034	-1.6991	-0.0346	-3.3641
0.035	-1.6891	-0.0356	-3.3346
0.036	-1.6792	-0.0367	-3.3059
0.037	-1.6695	-0.0377	-3.2780
0.038	-1.6599	-0.0387	-3.2508
0.039	-1.6505	-0.0398	-3.2243
0.040	-1.6413	-0.0408	-3.1985
0.041	-1.6322	-0.0419	-3.1733
0.042	-1.6232	-0.0429	-3.1487
0.043	-1.6144	-0.0440	-3.1246
0.044	-1.6057	-0.0450	-3.1011
0.045	-1.5971	-0.0460	-3.0781
0.046	-1.5886	-0.0471	-3.0556
0.047	-1.5803	-0.0481	-3.0336
0.048	-1.5720	-0.0492	-3.0120
0.049	-1.5639	-0.0502	-2.9909

Table T3 *Cont.*

F(.)	X_n	X_e	X_w
0.050	−1.5559	−0.0513	−2.9701
0.051	−1.5480	−0.0523	−2.9498
0.052	−1.5402	−0.0534	−2.9299
0.053	−1.5324	−0.0545	−2.9103
0.054	−1.5248	−0.0555	−2.8911
0.055	−1.5172	−0.0566	−2.8722
0.056	−1.5098	−0.0576	−2.8537
0.057	−1.5024	−0.0587	−2.8355
0.058	−1.4951	−0.0598	−2.8175
0.059	−1.4879	−0.0608	−2.7999
0.060	−1.4807	−0.0619	−2.7826
0.061	−1.4737	−0.0629	−2.7655
0.062	−1.4667	−0.0640	−2.7487
0.063	−1.4598	−0.0651	−2.7322
0.064	−1.4529	−0.0661	−2.7159
0.065	−1.4461	−0.0672	−2.6999
0.066	−1.4394	−0.0683	−2.6841
0.067	−1.4327	−0.0694	−2.6685
0.068	−1.4261	−0.0704	−2.6532
0.069	−1.4196	−0.0715	−2.6381
0.070	−1.4131	−0.0726	−2.6231
0.071	−1.4067	−0.0737	−2.6084
0.072	−1.4004	−0.0747	−2.5939
0.073	−1.3941	−0.0758	−2.5796
0.074	−1.3878	−0.0769	−2.5654
0.075	−1.3816	−0.0780	−2.5515
0.076	−1.3755	−0.0790	−2.5377
0.077	−1.3694	−0.0801	−2.5241
0.078	−1.3633	−0.0812	−2.5107
0.079	−1.3573	−0.0823	−2.4974
0.080	−1.3514	−0.0834	−2.4843
0.081	−1.3455	−0.0845	−2.4713
0.082	−1.3396	−0.0856	−2.4585
0.083	−1.3338	−0.0867	−2.4459
0.084	−1.3280	−0.0877	−2.4333
0.085	−1.3223	−0.0888	−2.4210
0.086	−1.3166	−0.0899	−2.4087
0.087	−1.3110	−0.0910	−2.3966
0.088	−1.3054	−0.0921	−2.3847
0.089	−1.2998	−0.0932	−2.3728
0.090	−1.2943	−0.0943	−2.3611
0.091	−1.2888	−0.0954	−2.3495
0.092	−1.2833	−0.0965	−2.3381
0.093	−1.2779	−0.0976	−2.3267
0.094	−1.2725	−0.0987	−2.3155
0.095	−1.2672	−0.0998	−2.3043
0.096	−1.2619	−0.1009	−2.2933
0.097	−1.2566	−0.1020	−2.2824
0.098	−1.2514	−0.1031	−2.2716

Table T3 *Cont.*

$F(.)$	X_n	X_e	X_w
0.099	−1.2462	−0.1043	−2.2609
0.100	−1.2410	−0.1054	−2.2503
0.101	−1.2359	−0.1065	−2.2398
0.102	−1.2307	−0.1076	−2.2294
0.103	−1.2257	−0.1087	−2.2191
0.104	−1.2206	−0.1098	−2.2089
0.105	−1.2156	−0.1109	−2.1988
0.106	−1.2106	−0.1121	−2.1888
0.107	−1.2056	−0.1132	−2.1788
0.108	−1.2007	−0.1143	−2.1690
0.109	−1.1958	−0.1154	−2.1592
0.110	−1.1909	−0.1165	−2.1495
0.111	−1.1861	−0.1177	−2.1399
0.112	−1.1812	−0.1188	−2.1304
0.113	−1.1764	−0.1199	−2.1210
0.114	−1.1717	−0.1210	−2.1116
0.115	−1.1669	−0.1222	−2.1023
0.116	−1.1622	−0.1233	−2.0931
0.117	−1.1575	−0.1244	−2.0840
0.118	−1.1528	−0.1256	−2.0749
0.119	−1.1482	−0.1267	−2.0659
0.120	−1.1436	−0.1278	−2.0570
0.121	−1.1390	−0.1290	−2.0481
0.122	−1.1344	−0.1301	−2.0393
0.123	−1.1298	−0.1313	−2.0306
0.124	−1.1253	−0.1324	−2.0220
0.125	−1.1208	−0.1335	−2.0134
0.126	−1.1163	−0.1347	−2.0048
0.127	−1.1118	−0.1358	−1.9964
0.128	−1.1074	−0.1370	−1.9880
0.129	−1.1029	−0.1381	−1.9796
0.130	−1.0985	−0.1393	−1.9713
0.131	−1.0941	−0.1404	−1.9631
0.132	−1.0898	−0.1416	−1.9550
0.133	−1.0854	−0.1427	−1.9468
0.134	−1.0811	−0.1439	−1.9388
0.135	−1.0768	−0.1450	−1.9308
0.136	−1.0725	−0.1462	−1.9228
0.137	−1.0682	−0.1473	−1.9150
0.138	−1.0640	−0.1485	−1.9071
0.139	−1.0597	−0.1497	−1.8993
0.140	−1.0555	−0.1508	−1.8916
0.141	−1.0513	−0.1520	−1.8839
0.142	−1.0471	−0.1532	−1.8763
0.143	−1.0430	−0.1543	−1.8687
0.144	−1.0388	−0.1555	−1.8612
0.145	−1.0347	−0.1567	−1.8537
0.146	−1.0306	−0.1578	−1.8462
0.147	−1.0265	−0.1590	−1.8388

Table T3 *Cont.*

$F(.)$	X_n	X_e	X_w
0.148	−1.0224	0.1602	−1.8315
0.149	−1.0184	−0.1614	−1.8242
0.150	−1.0143	−0.1625	−1.8169
0.151	−1.0103	−0.1637	−1.8097
0.152	−1.0063	−0.1649	−1.8025
0.153	−1.0023	−0.1661	−1.7954
0.154	−0.9983	−0.1672	−1.7883
0.155	−0.9943	−0.1684	−1.7813
0.156	−0.9903	−0.1696	−1.7742
0.157	−0.9864	−0.1708	−1.7673
0.158	−0.9825	−0.1720	−1.7604
0.159	−0.9786	−0.1732	−1.7535
0.160	−0.9747	−0.1744	−1.7466
0.161	−0.9708	−0.1756	−1.7398
0.162	−0.9669	−0.1767	−1.7330
0.163	−0.9630	−0.1779	−1.7263
0.164	−0.9592	−0.1791	−1.7196
0.165	−0.9554	−0.1803	−1.7130
0.166	−0.9516	−0.1815	−1.7063
0.167	−0.9477	−0.1827	−1.6997
0.168	−0.9440	−0.1839	−1.6932
0.169	−0.9402	−0.1851	−1.6867
0.170	−0.9364	−0.1863	−1.6802
0.171	−0.9327	−0.1875	−1.6737
0.172	−0.9289	−0.1888	−1.6673
0.173	−0.9252	−0.1900	−1.6609
0.174	−0.9215	−0.1912	−1.6546
0.175	−0.9178	−0.1924	−1.6483
0.176	−0.9141	−0.1936	−1.6420
0.177	−0.9104	−0.1948	−1.6357
0.178	−0.9067	−0.1960	−1.6295
0.179	−0.9031	−0.1972	−1.6233
0.180	−0.8994	−0.1985	−1.6172
0.181	−0.8958	−0.1997	−1.6110
0.182	−0.8922	−0.2009	−1.6049
0.183	−0.8885	−0.2021	−1.5989
0.184	−0.8849	−0.2034	−1.5928
0.185	−0.8814	−0.2046	−1.5868
0.186	−0.8778	−0.2058	−1.5808
0.187	−0.8742	−0.2070	−1.5749
0.188	−0.8706	−0.2083	−1.5689
0.189	−0.8671	−0.2095	−1.5630
0.190	−0.8636	−0.2107	−1.5572
0.191	−0.8600	−0.2120	−1.5513
0.192	−0.8565	−0.2132	−1.5455
0.193	−0.8530	−0.2144	−1.5397
0.194	−0.8495	−0.2157	−1.5339
0.195	−0.8460	−0.2169	−1.5282
0.196	−0.8425	−0.2182	−1.5225

Table T3 *Cont.*

$F(.)$	X_n	X_e	X_w
0.197	−0.8391	−0.2194	−1.5168
0.198	−0.8356	−0.2207	−1.5111
0.199	−0.8321	−0.2219	−1.5055
0.200	−0.8287	−0.2232	−1.4999
0.201	−0.8253	−0.2244	−1.4943
0.202	−0.8218	−0.2257	−1.4887
0.203	−0.8184	−0.2269	−1.4832
0.204	−0.8150	−0.2282	−1.4777
0.205	−0.8116	−0.2294	−1.4722
0.206	−0.8082	−0.2307	−1.4667
0.207	−0.8049	−0.2319	−1.4613
0.208	−0.8015	−0.2332	−1.4558
0.209	−0.7981	−0.2345	−1.4504
0.210	−0.7948	−0.2357	−1.4450
0.211	−0.7914	−0.2370	−1.4397
0.212	−0.7881	−0.2383	−1.4344
0.213	−0.7848	−0.2395	−1.4290
0.214	−0.7814	−0.2408	−1.4237
0.215	−0.7781	−0.2421	−1.4185
0.216	−0.7748	−0.2434	−1.4132
0.217	−0.7715	−0.2446	−1.4080
0.218	−0.7682	−0.2459	−1.4028
0.219	−0.7650	−0.2472	−1.3976
0.220	−0.7617	−0.2485	−1.3924
0.221	−0.7584	−0.2498	−1.3873
0.222	−0.7552	−0.2510	−1.3821
0.223	−0.7519	−0.2523	−1.3770
0.224	−0.7487	−0.2536	−1.3719
0.225	−0.7454	−0.2549	−1.3669
0.226	−0.7422	−0.2562	−1.3618
0.227	−0.7390	−0.2575	−1.3568
0.228	−0.7358	−0.2588	−1.3518
0.229	−0.7326	−0.2601	−1.3468
0.230	−0.7294	−0.2614	−1.3418
0.231	−0.7262	−0.2627	−1.3368
0.232	−0.7230	−0.2640	−1.3319
0.233	−0.7198	−0.2653	−1.3270
0.234	−0.7166	−0.2666	−1.3221
0.235	−0.7135	−0.2679	−1.3172
0.236	−0.7103	−0.2692	−1.3123
0.237	−0.7072	−0.2705	−1.3074
0.238	−0.7040	−0.2718	−1.3026
0.239	−0.7009	−0.2731	−1.2978
0.240	−0.6978	−0.2745	−1.2930
0.241	−0.6946	−0.2758	−1.2882
0.242	−0.6915	−0.2771	−1.2834
0.243	−0.6884	−0.2784	−1.2787
0.244	−0.6853	−0.2797	−1.2739
0.245	−0.6822	−0.2811	−1.2692

Table T3 *Cont.*

$F(.)$	X_n	X_e	X_w
0.246	−0.6791	−0.2824	−1.2645
0.247	−0.6760	−0.2837	−1.2598
0.248	−0.6729	−0.2850	−1.2551
0.249	−0.6698	−0.2864	−1.2505
0.250	−0.6668	−0.2877	−1.2458
0.251	−0.6637	−0.2890	−1.2412
0.252	−0.6607	−0.2904	−1.2366
0.253	−0.6576	−0.2917	−1.2320
0.254	−0.6546	−0.2930	−1.2274
0.255	−0.6515	−0.2944	−1.2229
0.256	−0.6485	−0.2957	−1.2183
0.257	−0.6454	−0.2971	−1.2138
0.258	−0.6424	−0.2984	−1.2092
0.259	−0.6394	−0.2998	−1.2047
0.260	−0.6364	−0.3011	−1.2002
0.261	−0.6334	−0.3025	−1.1958
0.262	−0.6304	−0.3038	−1.1913
0.263	−0.6274	−0.3052	−1.1868
0.264	−0.6244	−0.3065	−1.1824
0.265	−0.6214	−0.3079	−1.1780
0.266	−0.6184	−0.3093	−1.1736
0.267	−0.6154	−0.3106	−1.1692
0.268	−0.6125	−0.3120	−1.1648
0.269	−0.6095	−0.3134	−1.1604
0.270	−0.6065	−0.3147	−1.1560
0.271	−0.6036	−0.3161	−1.1517
0.272	−0.6006	−0.3175	−1.1474
0.273	−0.5977	−0.3188	−1.1430
0.274	−0.5947	−0.3202	−1.1387
0.275	−0.5918	−0.3216	−1.1344
0.276	−0.5889	−0.3230	−1.1302
0.277	−0.5860	−0.3244	−1.1259
0.278	−0.5830	−0.3257	−1.1216
0.279	−0.5801	−0.3271	−1.1174
0.280	−0.5772	−0.3285	−1.1132
0.281	−0.5743	−0.3299	−1.1089
0.282	−0.5714	−0.3313	−1.1047
0.283	−0.5685	−0.3327	−1.1005
0.284	−0.5656	−0.3341	−1.0963
0.285	−0.5627	−0.3355	−1.0922
0.286	−0.5598	−0.3369	−1.0880
0.287	−0.5569	−0.3383	−1.0838
0.288	−0.5540	−0.3397	−1.0797
0.289	−0.5512	−0.3411	−1.0756
0.290	−0.5483	−0.3425	−1.0715
0.291	−0.5454	−0.3439	−1.0673
0.292	−0.5426	−0.3453	−1.0633
0.293	−0.5397	−0.3467	−1.0592
0.294	−0.5369	−0.3482	−1.0551

Table T3 *Cont.*

$F(.)$	X_n	X_e	X_w
0.295	−0.5340	−0.3496	−1.0510
0.296	−0.5312	−0.3510	−1.0470
0.297	−0.5283	−0.3524	−1.0429
0.298	−0.5255	−0.3538	−1.0389
0.299	−0.5227	−0.3553	−1.0349
0.300	−0.5198	−0.3567	−1.0309
0.301	−0.5170	−0.3581	−1.0269
0.302	−0.5142	−0.3596	−1.0229
0.303	−0.5114	−0.3610	−1.0189
0.304	−0.5086	−0.3624	−1.0149
0.305	−0.5057	−0.3639	−1.0110
0.306	−0.5029	−0.3653	−1.0070
0.307	−0.5001	−0.3667	−1.0031
0.308	−0.4973	−0.3682	−0.9992
0.309	−0.4945	−0.3696	−0.9952
0.310	−0.4918	−0.3711	−0.9913
0.311	−0.4890	−0.3725	−0.9874
0.312	−0.4862	−0.3740	−0.9835
0.313	−0.4834	−0.3754	−0.9797
0.314	−0.4806	−0.3769	−0.9758
0.315	−0.4779	−0.3784	−0.9719
0.316	−0.4751	−0.3798	−0.9681
0.317	−0.4723	−0.3813	−0.9642
0.318	−0.4696	−0.3827	−0.9604
0.319	−0.4668	−0.3842	−0.9566
0.320	−0.4640	−0.3857	−0.9527
0.321	−0.4613	−0.3872	−0.9489
0.322	−0.4585	−0.3886	−0.9451
0.323	−0.4558	−0.3901	−0.9413
0.324	−0.4530	−0.3916	−0.9376
0.325	−0.4503	−0.3931	−0.9338
0.326	−0.4476	−0.3945	−0.9300
0.327	−0.4448	−0.3960	−0.9263
0.328	−0.4421	−0.3975	−0.9225
0.329	−0.4394	−0.3990	−0.9188
0.330	−0.4366	−0.4005	−0.9150
0.331	−0.4339	−0.4020	−0.9113
0.332	−0.4312	−0.4035	−0.9076
0.333	−0.4285	−0.4050	−0.9039
0.334	−0.4258	−0.4065	−0.9002
0.335	−0.4231	−0.4080	−0.8965
0.336	−0.4203	−0.4095	−0.8928
0.337	−0.4176	−0.4110	−0.8892
0.338	−0.4149	−0.4125	−0.8855
0.339	−0.4122	−0.4140	−0.8818
0.340	−0.4095	−0.4155	−0.8782
0.341	−0.4069	−0.4171	−0.8745
0.342	−0.4042	−0.4186	−0.8709
0.343	−0.4015	−0.4201	−0.8673

Table T3 *Cont.*

$F(.)$	X_n	X_e	X_w
0.344	0.3988	0.4216	−0.8637
0.345	−0.3961	−0.4231	−0.8600
0.346	−0.3934	−0.4247	−0.8564
0.347	−0.3907	−0.4262	−0.8528
0.348	−0.3881	−0.4277	−0.8492
0.349	−0.3854	−0.4293	−0.8457
0.350	−0.3827	−0.4308	−0.8421
0.351	−0.3801	−0.4323	−0.8385
0.352	−0.3774	−0.4339	−0.8350
0.353	−0.3747	−0.4354	−0.8314
0.354	−0.3721	−0.4370	−0.8279
0.355	−0.3694	−0.4385	−0.8243
0.356	−0.3668	−0.4401	−0.8208
0.357	−0.3641	−0.4416	−0.8173
0.358	−0.3615	−0.4432	−0.8137
0.359	−0.3588	−0.4448	−0.8102
0.360	−0.3562	−0.4463	−0.8067
0.361	−0.3535	−0.4479	−0.8032
0.362	−0.3509	−0.4494	−0.7997
0.363	−0.3482	−0.4510	−0.7963
0.364	−0.3456	−0.4526	−0.7928
0.365	−0.3430	−0.4542	−0.7893
0.366	−0.3403	−0.4557	−0.7858
0.367	−0.3377	−0.4573	−0.7824
0.368	−0.3351	−0.4589	−0.7789
0.369	−0.3325	−0.4605	−0.7755
0.370	−0.3298	−0.4621	−0.7721
0.371	−0.3272	−0.4637	−0.7686
0.372	−0.3246	−0.4652	−0.7652
0.373	−0.3220	−0.4668	−0.7618
0.374	−0.3194	−0.4684	−0.7584
0.375	−0.3167	−0.4700	−0.7550
0.376	−0.3141	−0.4716	−0.7516
0.377	−0.3115	−0.4732	−0.7482
0.378	−0.3089	−0.4748	−0.7448
0.379	−0.3063	−0.4765	−0.7414
0.380	−0.3037	−0.4781	−0.7380
0.381	−0.3011	−0.4797	−0.7346
0.382	−0.2985	−0.4813	−0.7313
0.383	−0.2959	−0.4829	−0.7279
0.384	−0.2933	−0.4845	−0.7246
0.385	−0.2907	−0.4862	−0.7212
0.386	−0.2881	−0.4878	−0.7179
0.387	−0.2855	−0.4894	−0.7145
0.388	−0.2829	−0.4911	−0.7112
0.389	−0.2804	−0.4927	−0.7079
0.390	−0.2778	−0.4943	−0.7046
0.391	−0.2752	−0.4960	−0.7012
0.392	−0.2726	−0.4976	−0.6979

Table T3 *Cont.*

$F(.)$	X_n	X_e	X_w
0.393	−0.2700	−0.4993	−0.6946
0.394	−0.2675	−0.5009	−0.6913
0.395	−0.2649	−0.5026	−0.6880
0.396	−0.2623	−0.5042	−0.6848
0.397	−0.2597	−0.5059	−0.6815
0.398	−0.2572	−0.5075	−0.6782
0.399	−0.2546	−0.5092	−0.6749
0.400	−0.2520	−0.5109	−0.6717
0.401	−0.2494	−0.5125	−0.6684
0.402	−0.2469	−0.5142	−0.6652
0.403	−0.2443	−0.5159	−0.6619
0.404	−0.2417	−0.5175	−0.6587
0.405	−0.2392	−0.5192	−0.6554
0.406	−0.2366	−0.5209	−0.6522
0.407	−0.2341	−0.5226	−0.6490
0.408	−0.2315	−0.5243	−0.6457
0.409	−0.2290	−0.5260	−0.6425
0.410	−0.2264	−0.5277	−0.6393
0.411	−0.2238	−0.5294	−0.6361
0.412	−0.2213	−0.5311	−0.6329
0.413	−0.2187	−0.5328	−0.6297
0.414	−0.2162	−0.5345	−0.6265
0.415	−0.2136	−0.5362	−0.6233
0.416	−0.2111	−0.5379	−0.6201
0.417	−0.2086	−0.5396	−0.6169
0.418	−0.2060	−0.5413	−0.6137
0.419	−0.2035	−0.5430	−0.6106
0.420	−0.2009	−0.5448	−0.6074
0.421	−0.1984	−0.5465	−0.6042
0.422	−0.1958	−0.5482	−0.6011
0.423	−0.1933	−0.5499	−0.5979
0.424	−0.1908	−0.5517	−0.5948
0.425	−0.1882	−0.5534	−0.5916
0.426	−0.1857	−0.5552	−0.5885
0.427	−0.1832	−0.5569	−0.5854
0.428	−0.1806	−0.5587	−0.5822
0.429	−0.1781	−0.5604	−0.5791
0.430	−0.1756	−0.5622	−0.5760
0.431	−0.1730	−0.5639	−0.5729
0.432	−0.1705	−0.5657	−0.5697
0.433	−0.1680	−0.5674	−0.5666
0.434	−0.1654	−0.5692	−0.5635
0.435	−0.1629	−0.5710	−0.5604
0.436	−0.1604	−0.5727	−0.5573
0.437	−0.1579	−0.5745	−0.5542
0.438	−0.1553	−0.5763	−0.5511
0.439	−0.1528	−0.5781	−0.5481
0.440	−0.1503	−0.5799	−0.5450
0.441	−0.1478	−0.5816	−0.5419

Table T3 *Cont.*

$F(.)$	X_n	X_e	X_w
0.442	-0.1453	-0.5834	-0.5388
0.443	-0.1427	-0.5852	-0.5358
0.444	-0.1402	-0.5870	-0.5327
0.445	-0.1377	-0.5888	-0.5296
0.446	-0.1352	-0.5906	-0.5266
0.447	-0.1327	-0.5924	-0.5235
0.448	-0.1301	-0.5942	-0.5205
0.449	-0.1276	-0.5961	-0.5174
0.450	-0.1251	-0.5979	-0.5144
0.451	-0.1226	-0.5997	-0.5113
0.452	-0.1201	-0.6015	-0.5083
0.453	-0.1176	-0.6033	-0.5053
0.454	-0.1151	-0.6052	-0.5022
0.455	-0.1126	-0.6070	-0.4992
0.456	-0.1100	-0.6088	-0.4962
0.457	-0.1075	-0.6107	-0.4932
0.458	-0.1050	-0.6125	-0.4902
0.459	-0.1025	-0.6144	-0.4871
0.460	-0.1000	-0.6162	-0.4841
0.461	-0.0975	-0.6181	-0.4811
0.462	-0.0950	-0.6199	-0.4781
0.463	-0.0925	-0.6218	-0.4751
0.464	-0.0900	-0.6237	-0.4721
0.465	-0.0875	-0.6255	-0.4692
0.466	-0.0850	-0.6274	-0.4662
0.467	-0.0825	-0.6293	-0.4632
0.468	-0.0800	-0.6312	-0.4602
0.469	-0.0775	-0.6330	-0.4572
0.470	-0.0750	-0.6349	-0.4543
0.471	-0.0724	-0.6368	-0.4513
0.472	-0.0699	-0.6387	-0.4483
0.473	-0.0674	-0.6406	-0.4454
0.474	-0.0649	-0.6425	-0.4424
0.475	-0.0624	-0.6444	-0.4394
0.476	-0.0599	-0.6463	-0.4365
0.477	-0.0574	-0.6482	-0.4335
0.478	-0.0549	-0.6501	-0.4306
0.479	-0.0524	-0.6520	-0.4276
0.480	-0.0499	-0.6540	-0.4247
0.481	-0.0474	-0.6559	-0.4218
0.482	-0.0449	-0.6578	-0.4188
0.483	-0.0424	-0.6598	-0.4159
0.484	-0.0399	-0.6617	-0.4130
0.485	-0.0374	-0.6636	-0.4100
0.486	-0.0349	-0.6656	-0.4071
0.487	-0.0324	-0.6675	-0.4042
0.488	-0.0299	-0.6695	-0.4013
0.489	-0.0274	-0.6714	-0.3983
0.490	-0.0249	-0.6734	-0.3954

Table T3 *Cont.*

$F(.)$	X_n	X_e	X_w
0.491	−0.0224	−0.6754	−0.3925
0.492	−0.0199	−0.6773	−0.3896
0.493	−0.0174	−0.6793	−0.3867
0.494	−0.0149	−0.6813	−0.3838
0.495	−0.0124	−0.6832	−0.3809
0.496	−0.0099	−0.6852	−0.3780
0.497	−0.0074	−0.6872	−0.3751
0.498	−0.0049	−0.6892	−0.3722
0.499	−0.0024	−0.6912	−0.3693
0.500	0.0001	−0.6932	−0.3664
0.501	0.0026	−0.6952	−0.3636
0.502	0.0051	−0.6972	−0.3607
0.503	0.0076	−0.6992	−0.3578
0.504	0.0101	−0.7012	−0.3549
0.505	0.0126	−0.7032	−0.3520
0.506	0.0151	−0.7053	−0.3492
0.507	0.0176	−0.7073	−0.3463
0.508	0.0201	−0.7093	−0.3434
0.509	0.0226	−0.7114	−0.3406
0.510	0.0251	−0.7134	−0.3377
0.511	0.0276	−0.7154	−0.3349
0.512	0.0301	−0.7175	−0.3320
0.513	0.0326	−0.7195	−0.3291
0.514	0.0351	−0.7216	−0.3263
0.515	0.0376	−0.7237	−0.3234
0.516	0.0401	−0.7257	−0.3206
0.517	0.0426	−0.7278	−0.3177
0.518	0.0451	−0.7299	−0.3149
0.519	0.0476	−0.7319	−0.3121
0.520	0.0501	−0.7340	−0.3092
0.521	0.0526	−0.7361	−0.3064
0.522	0.0551	−0.7382	−0.3035
0.523	0.0576	−0.7403	−0.3007
0.524	0.0601	−0.7424	−0.2979
0.525	0.0626	−0.7445	−0.2951
0.526	0.0651	−0.7466	−0.2922
0.527	0.0676	−0.7487	−0.2894
0.528	0.0701	−0.7508	−0.2866
0.529	0.0726	−0.7529	−0.2838
0.530	0.0751	−0.7551	−0.2809
0.531	0.0777	−0.7572	−0.2781
0.532	0.0802	−0.7593	−0.2753
0.533	0.0827	−0.7615	−0.2725
0.534	0.0852	−0.7636	−0.2697
0.535	0.0877	−0.7658	−0.2669
0.536	0.0902	−0.7679	−0.2641
0.537	0.0927	−0.7701	−0.2613
0.538	0.0952	−0.7722	−0.2585
0.539	0.0977	−0.7744	−0.2557

Table T3 *Cont.*

$F(.)$	X_n	X_e	X_w
0.540	0.1003	−0.7766	−0.2529
0.541	0.1028	−0.7788	−0.2501
0.542	0.1053	−0.7809	−0.2473
0.543	0.1078	−0.7831	−0.2445
0.544	0.1103	−0.7853	−0.2417
0.545	0.1128	−0.7875	−0.2389
0.546	0.1154	−0.7897	−0.2361
0.547	0.1179	−0.7919	−0.2333
0.548	0.1204	−0.7941	−0.2305
0.549	0.1229	−0.7963	−0.2277
0.550	0.1254	−0.7986	−0.2249
0.551	0.1280	−0.8008	−0.2222
0.552	0.1305	−0.8030	−0.2194
0.553	0.1330	−0.8053	−0.2166
0.554	0.1355	−0.8075	−0.2138
0.555	0.1381	−0.8097	−0.2110
0.556	0.1406	−0.8120	−0.2083
0.557	0.1431	−0.8142	−0.2055
0.558	0.1457	−0.8165	−0.2027
0.559	0.1482	−0.8188	−0.2000
0.560	0.1507	−0.8210	−0.1972
0.561	0.1533	−0.8233	−0.1944
0.562	0.1558	−0.8256	−0.1916
0.563	0.1583	−0.8279	−0.1889
0.564	0.1609	−0.8302	−0.1861
0.565	0.1634	−0.8325	−0.1834
0.566	0.1659	−0.8348	−0.1806
0.567	0.1685	−0.8371	−0.1778
0.568	0.1710	−0.8394	−0.1751
0.569	0.1736	−0.8417	−0.1723
0.570	0.1761	−0.8440	−0.1696
0.571	0.1786	−0.8464	−0.1668
0.572	0.1812	−0.8487	−0.1641
0.573	0.1837	−0.8510	−0.1613
0.574	0.1863	−0.8534	−0.1586
0.575	0.1888	−0.8557	−0.1558
0.576	0.1914	−0.8581	−0.1531
0.577	0.1939	−0.8604	−0.1503
0.578	0.1965	−0.8628	−0.1476
0.579	0.1990	−0.8652	−0.1448
0.580	0.2016	−0.8676	−0.1421
0.581	0.2041	−0.8699	−0.1393
0.582	0.2067	−0.8723	−0.1366
0.583	0.2093	−0.8747	−0.1338
0.584	0.2118	−0.8771	−0.1311
0.585	0.2144	−0.8795	−0.1284
0.586	0.2169	−0.8820	−0.1256
0.587	0.2195	−0.8844	−0.1229
0.588	0.2221	−0.8868	−0.1201

Table T3 *Cont.*

$F(.)$	X_n	X_e	X_w
0.589	0.2246	-0.8892	-0.1174
0.590	0.2272	-0.8917	-0.1147
0.591	0.2298	-0.8941	-0.1119
0.592	0.2324	-0.8966	-0.1092
0.593	0.2349	-0.8990	-0.1065
0.594	0.2375	-0.9015	-0.1037
0.595	0.2401	-0.9039	-0.1010
0.596	0.2427	-0.9064	-0.0983
0.597	0.2453	-0.9089	-0.0955
0.598	0.2478	-0.9114	-0.0928
0.599	0.2504	-0.9139	-0.0901
0.600	0.2530	-0.9164	-0.0873
0.601	0.2556	-0.9189	-0.0846
0.602	0.2582	-0.9214	-0.0819
0.603	0.2608	-0.9239	-0.0792
0.604	0.2634	-0.9264	-0.0764
0.605	0.2660	-0.9289	-0.0737
0.606	0.2686	-0.9315	-0.0710
0.607	0.2712	-0.9340	-0.0683
0.608	0.2738	-0.9366	-0.0655
0.609	0.2764	-0.9391	-0.0628
0.610	0.2790	-0.9417	-0.0601
0.611	0.2816	-0.9442	-0.0574
0.612	0.2842	-0.9468	-0.0546
0.613	0.2868	-0.9494	-0.0519
0.614	0.2894	-0.9520	-0.0492
0.615	0.2920	-0.9546	-0.0465
0.616	0.2946	-0.9572	-0.0438
0.617	0.2973	-0.9598	-0.0410
0.618	0.2999	-0.9624	-0.0383
0.619	0.3025	-0.9650	-0.0356
0.620	0.3051	-0.9677	-0.0329
0.621	0.3077	-0.9703	-0.0302
0.622	0.3104	-0.9729	-0.0274
0.623	0.3130	-0.9756	-0.0247
0.624	0.3156	-0.9782	-0.0220
0.625	0.3183	-0.9809	-0.0193
0.626	0.3209	-0.9836	-0.0166
0.627	0.3236	-0.9863	-0.0138
0.628	0.3262	-0.9889	-0.0111
0.629	0.3288	-0.9916	-0.0084
0.630	0.3315	-0.9943	-0.0057
0.631	0.3341	-0.9970	-0.0030
0.632	0.3368	-0.9998	-0.0002
0.633	0.3394	-1.0025	0.0025
0.634	0.3421	-1.0052	0.0052
0.635	0.3448	-1.0079	0.0079
0.636	0.3474	-1.0107	0.0106
0.637	0.3501	-1.0134	0.0133

Table T3 *Cont.*

$F(.)$	X_n	X_e	X_w
0.638	0.3528	−1.0162	0.0161
0.639	0.3554	−1.0190	0.0188
0.640	0.3581	−1.0217	0.0215
0.641	0.3608	−1.0245	0.0242
0.642	0.3634	−1.0273	0.0269
0.643	0.3661	−1.0301	0.0297
0.644	0.3688	−1.0329	0.0324
0.645	0.3715	−1.0357	0.0351
0.646	0.3742	−1.0385	0.0378
0.647	0.3769	−1.0414	0.0405
0.648	0.3796	−1.0442	0.0433
0.649	0.3823	−1.0471	0.0460
0.650	0.3850	−1.0499	0.0487
0.651	0.3877	−1.0528	0.0514
0.652	0.3904	−1.0556	0.0541
0.653	0.3931	−1.0585	0.0569
0.654	0.3958	−1.0614	0.0596
0.655	0.3985	−1.0643	0.0623
0.656	0.4012	−1.0672	0.0650
0.657	0.4039	−1.0701	0.0678
0.658	0.4067	−1.0730	0.0705
0.659	0.4094	−1.0760	0.0732
0.660	0.4121	−1.0789	0.0759
0.661	0.4148	−1.0818	0.0787
0.662	0.4176	−1.0848	0.0814
0.663	0.4203	−1.0878	0.0841
0.664	0.4231	−1.0907	0.0869
0.665	0.4258	−1.0937	0.0896
0.666	0.4285	−1.0967	0.0923
0.667	0.4313	−1.0997	0.0950
0.668	0.4340	−1.1027	0.0978
0.669	0.4368	−1.1057	0.1005
0.670	0.4396	−1.1088	0.1032
0.671	0.4423	−1.1118	0.1060
0.672	0.4451	−1.1148	0.1087
0.673	0.4479	−1.1179	0.1114
0.674	0.4506	−1.1210	0.1142
0.675	0.4534	−1.1240	0.1169
0.676	0.4562	−1.1271	0.1197
0.677	0.4590	−1.1302	0.1224
0.678	0.4618	−1.1333	0.1251
0.679	0.4646	−1.1364	0.1279
0.680	0.4674	−1.1395	0.1306
0.681	0.4702	−1.1427	0.1334
0.682	0.4730	−1.1458	0.1361
0.683	0.4758	−1.1490	0.1389
0.684	0.4786	−1.1521	0.1416
0.685	0.4814	−1.1553	0.1443
0.686	0.4842	−1.1585	0.1471

Table T3 *Cont.*

$F(.)$	X_n	X_e	X_w
0.687	0.4870	-1.1617	0.1498
0.688	0.4899	-1.1649	0.1526
0.689	0.4927	-1.1681	0.1553
0.690	0.4955	-1.1713	0.1581
0.691	0.4984	-1.1745	0.1609
0.692	0.5012	-1.1778	0.1636
0.693	0.5041	-1.1810	0.1664
0.694	0.5069	-1.1843	0.1691
0.695	0.5098	-1.1876	0.1719
0.696	0.5126	-1.1908	0.1747
0.697	0.5155	-1.1941	0.1774
0.698	0.5183	-1.1974	0.1802
0.699	0.5212	-1.2008	0.1829
0.700	0.5241	-1.2041	0.1857
0.701	0.5270	-1.2074	0.1885
0.702	0.5299	-1.2108	0.1913
0.703	0.5327	-1.2141	0.1940
0.704	0.5356	-1.2175	0.1968
0.705	0.5385	-1.2209	0.1996
0.706	0.5414	-1.2243	0.2024
0.707	0.5443	-1.2277	0.2051
0.708	0.5473	-1.2311	0.2079
0.709	0.5502	-1.2345	0.2107
0.710	0.5531	-1.2380	0.2135
0.711	0.5560	-1.2414	0.2163
0.712	0.5590	-1.2449	0.2191
0.713	0.5619	-1.2484	0.2219
0.714	0.5648	-1.2519	0.2246
0.715	0.5678	-1.2554	0.2274
0.716	0.5707	-1.2589	0.2302
0.717	0.5737	-1.2624	0.2330
0.718	0.5766	-1.2660	0.2358
0.719	0.5796	-1.2695	0.2386
0.720	0.5826	-1.2731	0.2414
0.721	0.5856	-1.2767	0.2443
0.722	0.5885	-1.2803	0.2471
0.723	0.5915	-1.2839	0.2499
0.724	0.5945	-1.2875	0.2527
0.725	0.5975	-1.2911	0.2555
0.726	0.6005	-1.2948	0.2583
0.727	0.6035	-1.2984	0.2611
0.728	0.6065	-1.3021	0.2640
0.729	0.6095	-1.3058	0.2668
0.730	0.6126	-1.3095	0.2696
0.731	0.6156	-1.3132	0.2724
0.732	0.6186	-1.3169	0.2753
0.733	0.6217	-1.3206	0.2781
0.734	0.6247	-1.3244	0.2810
0.735	0.6278	-1.3282	0.2838

Table T3 *Cont.*

$F(.)$	X_n	X_e	X_w
0.736	0.6308	-1.3319	0.2866
0.737	0.6339	-1.3357	0.2895
0.738	0.6370	-1.3395	0.2923
0.739	0.6400	-1.3434	0.2952
0.740	0.6431	-1.3472	0.2980
0.741	0.6462	-1.3511	0.3009
0.742	0.6493	-1.3549	0.3037
0.743	0.6524	-1.3588	0.3066
0.744	0.6555	-1.3627	0.3095
0.745	0.6586	-1.3666	0.3123
0.746	0.6618	-1.3706	0.3152
0.747	0.6649	-1.3745	0.3181
0.748	0.6680	-1.3785	0.3210
0.749	0.6712	-1.3824	0.3239
0.750	0.6743	-1.3864	0.3267
0.751	0.6775	-1.3904	0.3296
0.752	0.6806	-1.3945	0.3325
0.753	0.6838	-1.3985	0.3354
0.754	0.6870	-1.4026	0.3383
0.755	0.6901	-1.4066	0.3412
0.756	0.6933	-1.4107	0.3441
0.757	0.6965	-1.4148	0.3470
0.758	0.6997	-1.4190	0.3499
0.759	0.7029	-1.4231	0.3528
0.760	0.7061	-1.4273	0.3558
0.761	0.7094	-1.4314	0.3587
0.762	0.7126	-1.4356	0.3616
0.763	0.7158	-1.4398	0.3645
0.764	0.7191	-1.4441	0.3675
0.765	0.7223	-1.4483	0.3704
0.766	0.7256	-1.4526	0.3733
0.767	0.7289	-1.4569	0.3763
0.768	0.7321	-1.4612	0.3792
0.769	0.7354	-1.4655	0.3822
0.770	0.7387	-1.4698	0.3851
0.771	0.7420	-1.4742	0.3881
0.772	0.7453	-1.4786	0.3911
0.773	0.7486	-1.4830	0.3940
0.774	0.7520	-1.4874	0.3970
0.775	0.7553	-1.4918	0.4000
0.776	0.7586	-1.4963	0.4030
0.777	0.7620	-1.5007	0.4060
0.778	0.7653	-1.5052	0.4090
0.779	0.7687	-1.5098	0.4119
0.780	0.7721	-1.5143	0.4149
0.781	0.7755	-1.5188	0.4180
0.782	0.7789	-1.5234	0.4210
0.783	0.7823	-1.5280	0.4240
0.784	0.7857	-1.5326	0.4270

Table T3 *Cont.*

$F(.)$	X_n	X_e	X_w
0.785	0.7891	−1.5373	0.4300
0.786	0.7925	−1.5419	0.4330
0.787	0.7960	−1.5466	0.4361
0.788	0.7994	−1.5513	0.4391
0.789	0.8029	−1.5561	0.4422
0.790	0.8064	−1.5608	0.4452
0.791	0.8098	−1.5656	0.4483
0.792	0.8133	−1.5704	0.4513
0.793	0.8168	−1.5752	0.4544
0.794	0.8203	−1.5801	0.4575
0.795	0.8238	−1.5849	0.4605
0.796	0.8274	−1.5898	0.4636
0.797	0.8309	−1.5947	0.4667
0.798	0.8345	−1.5997	0.4698
0.799	0.8380	−1.6046	0.4729
0.800	0.8416	−1.6096	0.4760
0.801	0.8452	−1.6146	0.4791
0.802	0.8488	−1.6197	0.4822
0.803	0.8524	−1.6247	0.4853
0.804	0.8560	−1.6298	0.4885
0.805	0.8596	−1.6349	0.4916
0.806	0.8632	−1.6401	0.4947
0.807	0.8669	−1.6453	0.4979
0.808	0.8705	−1.6504	0.5010
0.809	0.8742	−1.6557	0.5042
0.810	0.8779	−1.6609	0.5074
0.811	0.8816	−1.6662	0.5105
0.812	0.8853	−1.6715	0.5137
0.813	0.8890	−1.6768	0.5169
0.814	0.8927	−1.6822	0.5201
0.815	0.8965	−1.6876	0.5233
0.816	0.9002	−1.6930	0.5265
0.817	0.9040	−1.6985	0.5297
0.818	0.9078	−1.7039	0.5329
0.819	0.9116	−1.7095	0.5362
0.820	0.9154	−1.7150	0.5394
0.821	0.9192	−1.7206	0.5427
0.822	0.9231	−1.7262	0.5459
0.823	0.9269	−1.7318	0.5492
0.824	0.9308	−1.7375	0.5524
0.825	0.9346	−1.7432	0.5557
0.826	0.9385	−1.7489	0.5590
0.827	0.9424	−1.7547	0.5623
0.828	0.9464	−1.7605	0.5656
0.829	0.9503	−1.7663	0.5689
0.830	0.9542	−1.7722	0.5722
0.831	0.9582	−1.7781	0.5755
0.832	0.9622	−1.7840	0.5789
0.833	0.9662	−1.7900	0.5822

Table T3 Cont.

$F(.)$	X_n	X_e	X_w
0.834	0.9702	-1.7960	0.5856
0.835	0.9742	-1.8020	0.5889
0.836	0.9782	-1.8081	0.5923
0.837	0.9823	-1.8142	0.5957
0.838	0.9864	-1.8204	0.5990
0.839	0.9905	-1.8266	0.6024
0.840	0.9946	-1.8328	0.6059
0.841	0.9987	-1.8391	0.6093
0.842	1.0028	-1.8454	0.6127
0.843	1.0070	-1.8517	0.6161
0.844	1.0112	-1.8581	0.6196
0.845	1.0154	-1.8646	0.6230
0.846	1.0196	-1.8710	0.6265
0.847	1.0238	-1.8776	0.6300
0.848	1.0280	-1.8841	0.6335
0.849	1.0323	-1.8907	0.6370
0.850	1.0366	-1.8974	0.6405
0.851	1.0409	-1.9041	0.6440
0.852	1.0452	-1.9108	0.6475
0.853	1.0496	-1.9176	0.6511
0.854	1.0539	-1.9244	0.6546
0.855	1.0583	-1.9313	0.6582
0.856	1.0627	-1.9382	0.6618
0.857	1.0671	-1.9452	0.6653
0.858	1.0716	-1.9522	0.6690
0.859	1.0760	-1.9593	0.6726
0.860	1.0805	-1.9664	0.6762
0.861	1.0850	-1.9735	0.6798
0.862	1.0896	-1.9808	0.6835
0.863	1.0941	-1.9880	0.6872
0.864	1.0987	-1.9954	0.6908
0.865	1.1033	-2.0028	0.6945
0.866	1.1079	-2.0102	0.6982
0.867	1.1126	-2.0177	0.7020
0.868	1.1172	-2.0252	0.7057
0.869	1.1219	-2.0328	0.7094
0.870	1.1266	-2.0405	0.7132
0.871	1.1314	-2.0482	0.7170
0.872	1.1361	-2.0560	0.7208
0.873	1.1409	-2.0639	0.7246
0.874	1.1458	-2.0718	0.7284
0.875	1.1506	-2.0797	0.7322
0.876	1.1555	-2.0878	0.7361
0.877	1.1604	-2.0959	0.7400
0.878	1.1653	-2.1040	0.7439
0.879	1.1703	-2.1123	0.7478
0.880	1.1753	-2.1206	0.7517
0.881	1.1803	-2.1289	0.7556
0.882	1.1853	-2.1374	0.7596

Table T3 *Cont.*

$F(.)$	X_n	X_e	X_w
0.883	1.1904	−2.1459	0.7636
0.884	1.1955	−2.1545	0.7676
0.885	1.2007	−2.1631	0.7716
0.886	1.2058	−2.1719	0.7756
0.887	1.2110	−2.1807	0.7796
0.888	1.2163	−2.1896	0.7837
0.889	1.2216	−2.1986	0.7878
0.890	1.2269	−2.2076	0.7919
0.891	1.2322	−2.2167	0.7960
0.892	1.2376	−2.2260	0.8002
0.893	1.2430	−2.2353	0.8044
0.894	1.2484	−2.2447	0.8086
0.895	1.2539	−2.2541	0.8128
0.896	1.2595	−2.2637	0.8170
0.897	1.2650	−2.2734	0.8213
0.898	1.2706	−2.2831	0.8256
0.899	1.2763	−2.2930	0.8299
0.900	1.2819	−2.3030	0.8342
0.901	1.2877	−2.3130	0.8386
0.902	1.2934	−2.3232	0.8429
0.903	1.2992	−2.3334	0.8473
0.904	1.3051	−2.3438	0.8518
0.905	1.3110	−2.3543	0.8562
0.906	1.3169	−2.3649	0.8607
0.907	1.3229	−2.3756	0.8652
0.908	1.3290	−2.3864	0.8698
0.909	1.3351	−2.3973	0.8743
0.910	1.3412	−2.4084	0.8789
0.911	1.3474	−2.4195	0.8836
0.912	1.3536	−2.4308	0.8882
0.913	1.3599	−2.4423	0.8929
0.914	1.3663	−2.4538	0.8977
0.915	1.3727	−2.4655	0.9024
0.916	1.3791	−2.4774	0.9072
0.917	1.3857	−2.4894	0.9120
0.918	1.3922	−2.5015	0.9169
0.919	1.3989	−2.5138	0.9218
0.920	1.4056	−2.5262	0.9267
0.921	1.4123	−2.5388	0.9317
0.922	1.4192	−2.5515	0.9367
0.923	1.4261	−2.5644	0.9417
0.924	1.4330	−2.5775	0.9468
0.925	1.4401	−2.5908	0.9520
0.926	1.4472	−2.6042	0.9571
0.927	1.4544	−2.6178	0.9623
0.928	1.4616	−2.6316	0.9676
0.929	1.4689	−2.6456	0.9729
0.930	1.4764	−2.6598	0.9782
0.931	1.4839	−2.6742	0.9836

Table T3 *Cont.*

$F(.)$	X_n	X_e	X_w
0.932	1.4914	-2.6888	0.9891
0.933	1.4991	-2.7036	0.9946
0.934	1.5069	-2.7187	1.0001
0.935	1.5147	-2.7339	1.0057
0.936	1.5226	-2.7495	1.0114
0.937	1.5307	-2.7652	1.0171
0.938	1.5388	-2.7812	1.0229
0.939	1.5471	-2.7975	1.0287
0.940	1.5554	-2.8140	1.0346
0.941	1.5639	-2.8308	1.0406
0.942	1.5724	-2.8480	1.0466
0.943	1.5811	-2.8654	1.0527
0.944	1.5899	-2.8831	1.0589
0.945	1.5989	-2.9011	1.0651
0.946	1.6079	-2.9195	1.0714
0.947	1.6171	-2.9382	1.0778
0.948	1.6265	-2.9572	1.0843
0.949	1.6359	-2.9767	1.0908
0.950	1.6456	-2.9965	1.0974
0.951	1.6554	-3.0167	1.1042
0.952	1.6653	-3.0373	1.1110
0.953	1.6754	-3.0584	1.1179
0.954	1.6857	-3.0799	1.1249
0.955	1.6962	-3.1019	1.1320
0.956	1.7068	-3.1244	1.1392
0.957	1.7177	-3.1474	1.1466
0.958	1.7287	-3.1710	1.1540
0.959	1.7400	-3.1951	1.1616
0.960	1.7515	-3.2198	1.1693
0.961	1.7632	-3.2452	1.1772
0.962	1.7752	-3.2712	1.1851
0.963	1.7875	-3.2978	1.1933
0.964	1.8000	-3.3253	1.2016
0.965	1.8128	-3.3535	1.2100
0.966	1.8259	-3.3825	1.2186
0.967	1.8393	-3.4124	1.2274
0.968	1.8531	-3.4432	1.2364
0.969	1.8672	-3.4750	1.2456
0.970	1.8818	-3.5078	1.2550
0.971	1.8967	-3.5417	1.2646
0.972	1.9120	-3.5769	1.2745
0.973	1.9279	-3.6133	1.2846
0.974	1.9442	-3.6511	1.2950
0.975	1.9610	-3.6904	1.3057
0.976	1.9785	-3.7313	1.3167
0.977	1.9965	-3.7739	1.3281
0.978	2.0152	-3.8184	1.3398
0.979	2.0347	-3.8650	1.3520
0.980	2.0550	-3.9139	1.3645

Table T3 *Cont.*

$F(.)$	X_n	X_e	X_w
0.981	2.0761	−3.9653	1.3776
0.982	2.0982	−4.0195	1.3911
0.983	2.1214	−4.0767	1.4053
0.984	2.1458	−4.1375	1.4201
0.985	2.1715	−4.2022	1.4356
0.986	2.1988	−4.2714	1.4519
0.987	2.2278	−4.3457	1.4692
0.988	2.2588	−4.4260	1.4875
0.989	2.2921	−4.5133	1.5070
0.990	2.3282	−4.6089	1.5280
0.991	2.3676	−4.7147	1.5507
0.992	2.4111	−4.8330	1.5755
0.993	2.4596	−4.9672	1.6029
0.994	2.5148	−5.1222	1.6336
0.995	2.5788	−5.3058	1.6688
0.996	2.6556	−5.5308	1.7103
0.997	2.7523	−5.8217	1.7616
0.998	2.8844	−6.2335	1.8299

Table T4 Gamma function

B	$E(X)/A$	$V(X)/A^2$	B	$E(X)/A$	$V(X)/A^2$
0.5	2.000	20.000	5.4	0.922	0.039
0.6	1.505	6.997	5.5	0.923	0.038
0.7	1.266	3.427	5.6	0.924	0.037
0.8	1.133	2.040	5.7	0.925	0.036
0.9	1.052	1.372	5.8	0.926	0.035
1.0	1.000	1.000	5.9	0.927	0.034
1.1	0.965	0.771	6.0	0.927	0.033
1.2	0.941	0.620	6.1	0.928	0.032
1.3	0.924	0.513	6.2	0.929	0.031
1.4	0.911	0.435	6.3	0.930	0.030
1.5	0.903	0.376	6.4	0.931	0.029
1.6	0.897	0.329	6.5	0.932	0.028
1.7	0.892	0.292	6.6	0.932	0.028
1.8	0.889	0.261	6.7	0.933	0.027
1.9	0.887	0.236	6.8	0.934	0.026
2.0	0.886	0.215	6.9	0.934	0.026
2.1	0.886	0.196	7.0	0.935	0.025
2.2	0.886	0.181	7.1	0.936	0.024
2.3	0.886	0.167	7.2	0.937	0.024
2.4	0.886	0.155	7.3	0.937	0.023
2.5	0.887	0.144	7.4	0.938	0.023
2.6	0.888	0.135	7.5	0.938	0.022
2.7	0.889	0.126	7.6	0.939	0.022
2.8	0.890	0.119	7.7	0.940	0.021
2.9	0.892	0.112	7.8	0.940	0.021
3.0	0.893	0.106	7.9	0.941	0.020
3.1	0.894	0.100	8.0	0.941	0.020
3.2	0.896	0.095	8.1	0.942	0.019
3.3	0.897	0.090	8.2	0.943	0.019
3.4	0.898	0.085	8.3	0.943	0.019
3.5	0.900	0.081	8.4	0.944	0.018
3.6	0.901	0.078	8.5	0.944	0.018
3.7	0.902	0.074	8.6	0.945	0.018
3.8	0.904	0.071	8.7	0.945	0.017
3.9	0.905	0.068	8.8	0.946	0.017
4.0	0.906	0.065	8.9	0.946	0.017
4.1	0.908	0.062	9.0	0.947	0.016
4.2	0.909	0.060	9.1	0.947	0.016
4.3	0.910	0.057	9.2	0.948	0.016
4.4	0.911	0.055	9.3	0.948	0.015
4.5	0.912	0.053	9.4	0.949	0.015
4.6	0.914	0.051	9.5	0.949	0.015
4.7	0.915	0.049	9.6	0.949	0.014
4.8	0.916	0.048	9.7	0.950	0.014
4.9	0.917	0.046	9.8	0.950	0.014
5.0	0.918	0.045	9.9	0.951	0.014
5.1	0.919	0.043	10.0	0.951	0.013
5.2	0.920	0.042	10.1	0.951	0.013
5.3	0.921	0.040	10.2	0.952	0.013

Table T4 *Cont.*

B	E(X)/A	V(X)/A²	B	E(X)/A	V(X)/A²
10.3	0.952	0.013	15.2	0.966	0.007
10.4	0.953	0.013	15.3	0.966	0.006
10.5	0.953	0.012	15.4	0.966	0.006
10.6	0.953	0.012	15.5	0.966	0.006
10.7	0.954	0.012	15.6	0.966	0.006
10.8	0.954	0.012	15.7	0.967	0.006
10.9	0.954	0.012	15.8	0.967	0.006
11.0	0.955	0.011	15.9	0.967	0.006
11.1	0.955	0.011	16.0	0.967	0.006
11.2	0.955	0.011	16.1	0.967	0.006
11.3	0.956	0.011	16.2	0.968	0.006
11.4	0.956	0.011	16.3	0.968	0.006
11.5	0.956	0.011	16.4	0.968	0.006
11.6	0.957	0.010	16.5	0.968	0.006
11.7	0.957	0.010	16.6	0.968	0.006
11.8	0.957	0.010	16.7	0.968	0.006
11.9	0.958	0.010	16.8	0.969	0.005
12.0	0.958	0.010	16.9	0.969	0.005
12.1	0.958	0.010	17.0	0.969	0.005
12.2	0.959	0.010	17.1	0.969	0.005
12.3	0.959	0.009	17.2	0.969	0.005
12.4	0.959	0.009	17.3	0.969	0.005
12.5	0.959	0.009	17.4	0.970	0.005
12.6	0.960	0.009	17.5	0.970	0.005
12.7	0.960	0.009	17.6	0.970	0.005
12.8	0.960	0.009	17.7	0.970	0.005
12.9	0.960	0.009	17.8	0.970	0.005
13.0	0.961	0.009	17.9	0.970	0.005
13.1	0.961	0.008	18.0	0.970	0.005
13.2	0.961	0.008	18.1	0.971	0.005
13.3	0.961	0.008	18.2	0.971	0.005
13.4	0.962	0.008	18.3	0.971	0.005
13.5	0.962	0.008	18.4	0.971	0.005
13.6	0.962	0.008	18.5	0.971	0.005
13.7	0.962	0.008	18.6	0.971	0.005
13.8	0.963	0.008	18.7	0.971	0.005
13.9	0.963	0.008	18.8	0.972	0.005
14.0	0.963	0.008	18.9	0.972	0.004
14.1	0.963	0.007	19.0	0.972	0.004
14.2	0.964	0.007	19.1	0.972	0.004
14.3	0.964	0.007	19.2	0.972	0.004
14.4	0.964	0.007	19.3	0.972	0.004
14.5	0.964	0.007	19.4	0.972	0.004
14.6	0.964	0.007	19.5	0.972	0.004
14.7	0.965	0.007	19.6	0.973	0.004
14.8	0.965	0.007	19.7	0.973	0.004
14.9	0.965	0.007	19.8	0.973	0.004
15.0	0.965	0.007	19.9	0.973	0.004
15.1	0.966	0.007	20.0	0.973	0.004

Table T5 Kolmogorov–Smirnov test

n	dp p=0.90	p=0.95	p=0.99	n	dp p−0.90	p=0.95	p=0.99
1	0.9500	0.9750	0.9950	25	0.2377	0.2640	0.3166
2	0.7764	0.8419	0.9293	26	0.2332	0.2591	0.3106
3	0.6360	0.7076	0.8290	27	0.2290	0.2544	0.3050
4	0.5652	0.6239	0.7342	28	0.2250	0.2499	0.2997
5	0.5094	0.5633	0.6685	29	0.2212	0.2457	0.2947
6	0.4680	0.5193	0.6166	30	0.2176	0.2417	0.2899
7	0.4361	0.4834	0.5758	31	0.2141	0.2379	0.2853
8	0.4096	0.4543	0.5418	32	0.2108	0.2342	0.2809
9	0.3875	0.4300	0.5133	33	0.2077	0.2308	0.2768
10	0.3687	0.4092	0.4889	34	0.2047	0.2274	0.2728
11	0.3524	0.3912	0.4677	35	0.2018	0.2242	0.2690
12	0.3382	0.3754	0.4490	36	0.1991	0.2212	0.2653
13	0.3255	0.3614	0.4325	37	0.1965	0.2183	0.2618
14	0.3142	0.3489	0.4176	38	0.1939	0.2154	0.2584
15	0.3040	0.3376	0.4042	39	0.1915	0.2127	0.2552
16	0.2947	0.3273	0.3920	40	0.1891	0.2101	0.2521
17	0.2863	0.3180	0.3809	45	0.1786	0.1984	0.2380
18	0.2785	0.3094	0.3706	50	0.1696	0.1884	0.2260
19	0.2714	0.3014	0.3612	60	0.1551	0.1723	0.2067
20	0.2647	0.2941	0.3524	70	0.1438	0.1587	0.2521
21	0.2586	0.2872	0.3443	80	0.1347	0.1496	0.1795
22	0.2528	0.2809	0.3367	90	0.1271	0.1412	0.1694
23	0.2475	0.2749	0.3295	100	0.1207	0.1340	0.1608
24	0.2424	0.2693	0.3229				

REFERENCES AND FURTHER READING

Abramovitz, M., and A.I. Stegun: "Handbook of Mathematical Functions," Dover Publications, New York, 1964.

Barlow, R.E., and F. Proschan: "Mathematical Theory of Reliability," John Wiley, New York, 1965.

Bazovsky, I.: "Reliability Theory and Practices," Prentice-Hall, Englewood Cliffs, N.J., 1961.

Blanchard, B.S.: "Logistics Engineering and Management," 3rd ed., Prentice-Hall, Englewood Cliffs, N.J., 1986.

Blanchard, B.S., and W.J. Fabrycky: "Systems Engineering and Analysis," Prentice-Hall, Englewood Cliffs, N.J., 1981.

Blanchard, B. and E. Lowery: "Maintainability," McGraw-Hill, New York, 1969.

Carter, A.D.S.: "Mechanical Reliability," Macmillan, London, 1972.

Clarke, A.B., and R.L. Disney: "Probability and Random Processes for Engineers and Scientists," John Wiley, New York, 1970.

Gnedendko, B.V., Belyayev Yu, K. and A.D. Solovyev: "Mathematical Methods of Reliability Theory," Academic Press, London, 1970.

Haugen, E.B.: "Probabilistic Mechanical Design," The University of Arizona, Tucson, Arizona, 1972.

Haugen, E.B.: "Probabilistic Approach to Design," John Wiley, New York, 1968.

Hays, W.L., and R.L. Winkler: "Statistics: Probability, Inference, and Decision," Holt, Rinehart and Winston, Inc., New York, 1971.

Kapur, K.C., and L.R. Lamberson: "Reliability in Engineering Design," John Wiley, New York, 1977.

Knezevic, J.: "Reliability and Maintenance, Fundamentals," Lecture Notes, School of Engineering, University of Exeter, 1990.

Knezevic, J.: "Condition Parameter Based Approach to Calculation of Reliability Characteristics," *Reliability Engineering*, No. 19, pp. 29–39, Elsevier Applied Science Publishers Ltd., England, 1987.

Knezevic, J.: On the Application of a Condition Parameter Based Reliability Approach to Pipeline System, Proc. 2nd International Conference on Pipes, Pipeline and Pipeline Systems, Utrecht, Holland, June 1987.

Knezevic, J.: "Reliability of Engineering Systems," Lecture Notes, University of Exeter, Library of the Department of Engineering Science, 1983.

Kolmogorov, A.N.: "Foundation of the Theory of Probability," Chelsea Publishing Company, New York, 1950.

Kozlov, B.A., and I.A. Ushakov: "Reliability Handbook," Holt, Rinehart and Winston, New York, 1970.

Pronikov, A.S.: "Dependability and Durability of Engineering Products," Butterworths, London, 1973.

Roberts, N.: "Mathematical Methods in Reliability Engineering," McGraw-Hill, New York, 1964.

Treson, W.G.: "Reliability Handbook," McGraw-Hill, New York, 1966.

Von Alven, W.H. (Ed.): "Reliability Engineering," Prentice-Hall, Englewood Cliffs, N.J., 1964.

INDEX